INTRODUCTION TO
MICROSYSTEM PACKAGING TECHNOLOGY

INTRODUCTION TO MICROSYSTEM PACKAGING TECHNOLOGY

YUFENG JIN
ZHIPING WANG
JING CHEN

科学出版社
Science Press

CRC Press
Taylor & Francis Group
Boca Raton London New York

CRC Press is an imprint of the
Taylor & Francis Group, an **informa** business

CRC Press
Taylor & Francis Group
6000 Broken Sound Parkway NW, Suite 300
Boca Raton, FL 33487-2742

First issued in paperback 2018

© 2011 by Science Press
CRC Press is an imprint of Taylor & Francis Group, an Informa business

ISBN-13: 978-1-4398-1910-4 (hbk)
ISBN-13: 978-1-138-37425-6 (pbk)

Library of Congress Cataloging-in-Publication Data

Jin, Yufeng.
    Introduction to microsystem packaging technology / authors, Yufeng Jin, Zhiping Wang, Jing Chen.
        p. cm.
    "A CRC title."
    Includes bibliographical references and index.
    ISBN 978-1-4398-1910-4 (hardcover : alk. paper)
    1. Microelectronic packaging. I. Wang, Zhiping, 1962- Oct. 6 II. Chen, Jing, 1974- III. Title.

TK7870.15.J56 2010
621.381'046--dc22

                                       2009045580

Visit the Taylor & Francis Web site at
http://www.taylorandfrancis.com

and the CRC Press Web site at
http://www.crcpress.com

Foreword

There will be no terminus for the development of science and technology, and no stop to our endeavors. Half a century has past since the integrated circuit (IC) was invented. IC has been widely used in various areas as a basic unit of information processing and remains a cornerstone of information society. It has exceeded the expectations of its inventor in the width and depth of its development.

With the development of informatization, IC has stepped into the age of system on chip (SOC). SOC refers to the integration of processing, transmission, and storage functions into a chip, which might be called integrated microsystem in a narrow sense. In a broader sense the integrated system should contain information acquisition and servo implementation. Far beyond the domain of the electronics system, the microsystem in a broad sense is an emerging technological subject integrating microelectronics, precision mechanics, optoelectronics, and even biomedicine and hydrodynamics. Its foundation involves various subjects like physics, chemistry, mechanics, optics, biology, system, and control, while its application field is much wider than the IC. With its features of interdisciplinarity and possibility of application in multiple fields, many scientific and technological workers from various fields have been attracted to devote themselves to it. The 20 years since the mid 1980s has witnessed its tremendous development, and the 21st century will witness a more tremendous development. Besides difficulties in its design and processing techniques, packaging remains a striking problem for microsystem technology. Because packaging cannot be substituted with general integrated circuit encapsulation, the packaging problem has not been overcome. Packaging is a weak point especially in China; more exactly, it is a bottleneck for the development of microsystems.

Dr. Yufeng Jin, a young academic leader with a sound knowledge foundation and abundant practical experience, made multiple achievements including patents in the field of micro-electro-mechanical system (MEMS) while he worked at Peking University, as well as in Singapore. Now he succeeds me as director of the National Key Laboratory of Science and Technology on Micro/Nanometer Fabrication. I believe that the laboratory, under Dr. Jin's leadership, will make new breakthroughs and further contribute to the development of China's MEMS technology.

Introduction to Microsystem Packaging Technology, coauthored by Dr. Yufeng Jin, Dr. Zhiping Wang, and Dr. Jing Chen, involves various technologies in relation to microsystem packaging including packaging design, film materials and techniques, substrate, interconnect, encapsulation and sealing, device level package, MEMS packaging, modular assembly and optoelectronics packaging, system-in-package (SIP), inspection, reliability design, etc. This book can serve not only as teaching material and a reference book for graduates and senior undergraduates but also as a reference book for teachers and engineering technicians engaging in teaching and research in the field of MEMS. This book will definitely advance and be a breakthrough for China's microsystems packaging technology. I am very delighted that this monograph is to be published, and that the young generation of academic leaders are growing up.

Upon the publication of this book, I write this note to congratulate the authors on a job well done.

Yangyuan Wang

Lang Run Yuan, Beijing University

Preface

A microsystem is a kind of microminiaturized integrated system based on microelectronics, optoelectronics, RF (radio frequeucy) and wireless, micro-electro-mechanical system (MEMS), and related packaging and assembling technologies. Since the mid-1990s when microsystem-related industries pulled ahead of the steel and automotive industry, it has topped the pillar of industries, with its production value up to a trillion dollars. The microsystems packaging, with a production value up to a hundred billion dollars, is a significant technical link in the chain of information industry, a technical foundation upon which function chips are updated to application systems and an important technical guarantee for increase of information systems in technical content and value. The packaging industry, represented by microelectronics packaging technology, is playing more and more an important role in information technology, and together with design and manufacturing technologies, forms the foundation of the information industry.

Research on microsystems was initiated as early as the 1970s, but no real advance was made until the mid and late 1980s. From then on, it has developed into a research hotspot of advanced manufacturing technology.

Microsystems packaging technology chiefly includes material, substrate, interconnect, design, fabrication, test, reliability and system integration, and from the packaging aspect includes chip level packaging, device level packaging, module level packaging and complicated system level packaging. This book is written so that each chapter stands alone, some research examples added to present the information in a systematic and complete fashion. Each chapter provides targeted questions for reference in teaching and reading.

In order to impart the latest information on development of microsystems technology and help students learn about development trends in microsystems packaging, Dr. Yufeng Jin began to teach a course on basic knowledge of microsystems packaging to some graduates at the Shenzhen Graduate School and Institute of Microelectronics of Beijing University in the autumn of 2004. Coauthored by Dr. Yufeng Jin, Dr. Zhiping Wang, Dr. Jing Chen, and Dr. Jinwen Zhang, the first edition of this book in Chinese developed from the teaching material for the course combined with related scientific research of the authors, with reference to numerous documents. Designed for students majoring in microelectronics as well as related research, this book gives an overview of the latest advance of microsystems packaging technology and also of related multifaceted professional knowledge in a concise yet systematic way, providing an opportuing for people engaged in other research fields than packaging to broaden their knowledge.

This book consists of 10 chapters. Dr. Yufeng Jin wrote chapters 1, 2, 3, and 8; Dr. Zhiping Wang wrote chapters 7 and 10; Dr. Jing Chen wrote chapters 5 and 6; and Dr. Yufeng Jin together with Dr. Zhiping Wang wrote chapters 4 and 9; Dr. Yufeng Jin reviewed the whole book.

Our sincere thanks go to Dr. Min Miao, Professor Wengang Wu, and the students, including Ying Zhao, Yong Sun, Qingyu Sun, Qinghua Chen, and Mei Yang from Beijing University, and the students Xia Lou, Ran Fan, Dongsheng Xu, Qifang Hu, Jiaxun Zhang,

Ying Xie, Huiping Zhou, Haijing Wei, Lei Zhang, Tao Tao, Zhengpeng Zhu, etc., from Shenzhen Graduate School of Beijing University for their helpful work. Sincere thanks are also due to Professor Guoying Wu for his partial review of the manuscript. We also would like to thank Professor Xing Zhang from the School of Electronics Engineering and Computer Science of Beijing University, and Professor Tianyi Zhang from Shenzhen Graduate School of Beijing University, for their kind encouragement and support. We are especially grateful to Prof. Yangyuan Wang, an academician of the Chinese Academy of Sciences, for his instructional advice during the writing of the book, as well as for writing the foreword for the book despite his busy schedule.

Writing the second edition of this textbook in English is a great challenge for us because we are facing a situation of rapid development in microsystem packaging technology. A special thanks is extended to the Science Press in Beijing inviting us to edit this book, which allowed us the opportunity to examine and revise the content of the first edition. We have attempted to extract updated knowledge related to the research progresses of advanced packaging technology for microsystems.

The authors would like to express gratitude to Dr. Yandong He and Dr. Fei Su in revising Chapter 9, Dr. Min Miao in revising chapters 2 and 8, Dr. Peng Jin in revising Chapter 7, Dr. Yangfei Zhang and Dr. Wei Wang in updating Chapter 3, and Mr. Yunhui Zhu in updating Chapter 4. We would also like to recognize the excellent reviews of Dr. Daniel Shi from ASTRI, Hong Kong and Dr. Quoqi Zhang from NXP, the Netherlands, who helped us make important revisions in chapters 8, 9, and 10 in this book.

Also, it is possible that there are omissions or errors due to finite capability of the authors and tight writing time. Readers are invited to contact us with criticisms and corrections.

Yufeng Jin, Ph.D., Zhiping Wang, Ph.D., and Jing Chen, Ph.D.

Contents

CHAPTER 1

Introduction

1.1 What Are Microsystems?[1]

Microsystems may be defined differently by people in different disciplines. The microsystems discussed here are the miniaturized functional devices that are based on microelectronics technology, radio frequency (RF) and wireless, optical (or optical electronics), and micro-electro-mechanical systems (MEMS) technologies, and are manufactured, assembled, and integrated on carriers such as a leadframe or a substrate through encapsulation, interconnection, and other microfabrication techniques. As widely applied in the domain of information technology, the microsystems can also be called information technology microsystems. These microsystems consist of consumable electronic products such as calculators, personal computers, cell phones and video products, and information products covering computing products, telecommunication products, automobiles, avionics, and medical electronics. Today's activities and technical achievements are closely associated with these various integrated multifunctional microsystems.

The relationship between microsystem products and their related technologies, including microelectronics, optical electronics, RF, MEMS and packaging technology can be illustrated in Figure 1.1 as follows.

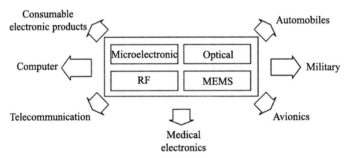

Figure 1.1   Relationship between microsystem products and their related technologies[1]

The vigorous development of microsystems is largely attributed to microelectronics achievements, since the invention and commercialization of the very large scale integrated circuit (VLSI) provided manufacturing platforms for miniaturized integrated systems.

Micromanufacturing is one of the essential technologies in microsystems. The microfabrication techniques in microelectronics are only part of the basic micromanufacturing techniques, which are determined by the structure difference between microsystems and the microelectronics circuit. A microsystem may include some movable parts, sensing units, and so on, and is usually a complex three-dimensional 3D multifunctional system such as an opto-electro-mechanical system. Integrated circuit manufacturing techniques are mainly two-dimensional or subsurface fabrication on silicon. Therefore, the whole manufacturing process of microsystems, which includes chip fabrication, integration and assembly, and packaging and testing, is much more complicated than those of integrated circuits.

The development of microelectronics technology enables the realization of many electronic products, such as high-speed information processors, large capacity memory, and products with ultralow power consumption. It is expected that future microsystem products will cover every aspect of our lives. For example, a large number of new products such as intelligent watches with integrated telephone and radio-recorder, multifunctional multimedia personal communication modules, micromedical implants, and minirobots will continuously turn up to meet people's demands for high-quality life.

1.2   Related Fundamentals of Microsystems

1.2.1   Microelectronics Technology

Microelectronics technology is used to manufacture miniaturized electronic components and circuits based on silicon materials through micron-level processing techniques. Through design and processing with a series of specific techniques, information systems that embody such functions as information gathering, processing, computing, transferring, storing, and execution, are integrated and embedded on silicon chips. Microelectronics is one of the most important fundamental technologies of modern industry, bringing revolutionary change to every walk of life. Its characteristics include:

(1) Strong technology penetration and high value added

When processed from raw material to products of microelectronic component and circuits, the added value increases tremendously.

(2) Sensitive market and short product cycle

Integrated circuits' development follows Moore's Law, that is, integration level increasing two times about every 18 months, and feature size decreasing 1.4 times every 3 years.

(3) Intensive technology and high information density

Microelectronics, including its related design, manufacturing, and packaging technologies, is an integration of up-to-date scientific and technological achievements of material, device physics, computer, optics, chemistry, vacuum, precision mechanics, physical and chemical analysis, and so forth.

In 2004, the global semiconductor market reached up to US$ 213 billion. In the first quarter of 2005, the global sales of semiconductors amounted to US$ 54.66 billion, a 14.4% increase over the same period in the previous year, of which Mainland China shared RMB (Renminbi) 77.32 billion. For the Asian-Pacific region, with China as its major part, the first quarter of 2005 saw a remarkable increase rate of 21.9%, the fastest one around the world.

1.2.2   RF and Wireless

Radio and wireless frequency usually ranges from 10 kHz to 1000 GHz. Their evolution can be traced back to the initial wireless transmission experiment made by Marconi in 1901. Wireless transmission frees people from wires and cables completely, and allows communication and information processing anytime and anywhere. Miniaturized, multifunctional, portable and low-cost RF devices have become the key for promotion and application of wireless communication as well as the important application domain of microsystems technology. At the same time, many of the latest noncommunication microsystems, such as advanced RF modules for automobile crash-resistance systems, global positioning system (GPS) modules, etc. have increasingly won the flavor of the market.

1.2.3   Optical Electronics

Optical technology includes acquiring, processing, transmission, and display of optical signals. Optoelectronic technology is an important member of optical technology. As we move into an information society featuring internet linkage and mass data processing, in order to break through the bottleneck of traditional means of signal transmission, optical fiber technology is developing rapidly. The optical fiber technology available now realizes

10 Tbit/s speed for data transmission on a single-mode optical fiber. With wavelength division multiplexing (WDM) technology, the capability of optical signal transmission can be improved greatly. Optical signals with different wavelengths (different colors) being transmitted through the same cable has also been realized. Optical fibers and cables can be applied not only to the trunk communication network, but also to the urban network, community networks, and even household access, which greatly changes life style and improves quality of life of mankind.

1.2.4 Microelectro-mechanical Systems

Micro-electro-mechanical systems (MEMS) are miniaturized devices or systems that usually consist of micromechanisms, microsensors, microactuators, and signal-processing circuit and control circuit, various interfaces, communications, and power. The all-in-one integration of electronic systems and non electronic systems realizes many functions previously unattainable so as to promote the harmonious development of acquiring, processing and handling information. Unlike microelectronics which focuses on two-dimensional fabrication techniques, MEMS lays stress on manufacturing of three-dimensional structure.

MEMS technology uses current microfabrication technology for mass production to produce multifunctional microsystems that are as small as integrated circuits in size. The present MEMS technology possesses the following characteristics:

(1) Diversification of research areas: the research of MEMS technology grows diversified, involving a variety of principles and areas, for example, various microsensors and microactuators such as microinertial devices, RF MEMS, microoptical devices, Bio MEMS, data storage, and so on.

(2) Diversification of manufacturing processes: for example, the combination of the traditional silicon bulk micromachining process, surface scarified layer process, silicon-fusion process, combination of deep reactive ion etching trench and bonding; Single Crystal Reactive Etching And Metallization (SCREAM), Lithographie Galvanoformaung Abformung (LIGA) and Ultraviolet-LIGA (UV-LIGA) processing, the metal sacrificial layer process combining thick photoresistance and electroplating technique; the combination of silicon bulk micromachining and surface sacrificed layer process.

(3) Integration of MEMS sensor chips and signal processing circuits in a single chip.

(4) Combined consideration of manufacturing and packaging MEMS chips.

1.3 What Is Microsystems Packaging?[1,2]

Microsystems packaging refers to the techniques with which one or more functional chips or elements are integrated or assembled into application products or modules. Microsystems packaging includes microelectronics packaging, optoelectronics packaging, RF packaging, MEMS packaging, and other multifunction system packaging.

"Packaging" is still very young as a technical term in electronics industry. In the age of vacuum tubes, fixing electron tubes on tube socket to form circuit was generally called "assembly" and no such a term concept as "packaging" existed then. The concept of Microsystems Packaging made its first appearance in microelectronics technology. With the invention of transistors and integrated circuits, protection of the chips was required due to their small size, fragilitys and diversity in functions and specifications. Reliable electrical connections with external circuits became essential with efficient mechanical protection and insulation. The packaging technology thus emerged as the times required.

Thus far, no official classification method is available for microsystems packaging. It is usually classified in the community by packaging materials, applications or packaging types, etc. In terms of architecture, microsystems packaging includes several hierarchies: chip-level package (or zero-level package), device-level package (or first-level package), board-level package (or second-level package board-level assembly), and motherboard-level or system level package (or third-level package system package). Figure 1.2 shows the packaging hierarchies in a typical microsystems packaging.

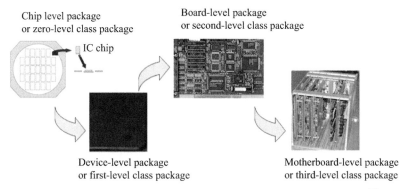

Figure 1.2   Packaging hierarchies in a typical microsystems packaging[1]

Technically, microsystems packaging involves three kinds of technologies: base chip technology, i.e., microelectronics chips, optoelectronics chip, MEMS chip, RF chip, etc., system engineering technology, and system packaging technology. The "chips" are functional elements of various kinds, and "system" all the object products, while "packaging" refers to connecting "chips" and other devices to system substrate to form object products. As for design and manufacturing technology of the "chips", which has been specifically discussed in other publications, they are beyond the scope of this book and hence are not included.

Figure 1.3 illustrates the chips, systems, and packaging mentioned above, and their relations. The area where chips and packaging overlap denotes the "packaged devices" which refers to the chips being device-level packaged. Integrated Circuit (IC) products such as microprocessors and memory modules in computers are examples of "packaged devices". The area where packaging and system overlap usually contains no "brain" (chip or device), thus not having the board-level system with functions of controlling and processing (also called system board). The area where chips and system overlap denotes subproduct, to which the currently booming system on chip (SOC) is a typical example. These subproducts usually have partial functions of a system, while generally requiring no complicated packaging techniques. They are designed mainly to meet the need of various system applications, with emphasis on functions and level of functional integration on a chip instead of packaging.

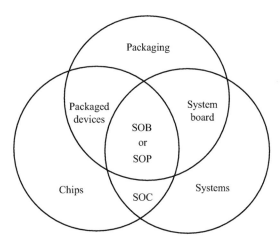

Figure 1.3   Relations of chips, systems, and packaging in microsystems packaging

It is expected that system on board (SOB) will remain the mainstream technology in electronics products in the near future since a large percentage of products are with a number of packaged devices and other components assembled on circuit board. New technologies, such as system on package (SOP) or system in package (SIP), will have unique applications where high functional density is required. Similar to SOC, SOP refers to packaging multifunctional chips in one module to provide all functions required, which may include analog, digital, light, RF and MEMS.

1.4   What Is Microelectronics Packaging?[2−4]

Microelectronics packaging, which accounts for 90% of the microsystems packaging market share, is an important and fundamental part of microsystems packaging technology. It is also an important part of microelectronics technology in general. This technology demands provision of mechanical protection, power supply, cooling, and electric and mechanical connections of microelectronics chips and elements to the outside world while minimizing the influence on their electrical performance.

The design and simulation is followed by the fabrication and application specific development in terms of technology chain of microelectronics. The whole fabrication stage can be divided into three phases, as shown in Figure 1.4. Within the domain of semiconductor device-level, the first two phases are called "front-end engineering" and "back-end engineering" where "front-" and "back-" are demarcated by the phase of wafer being diced into chips. Electronics packaging includes the two manufacturing steps in the square of the dashed lines in Figure 1.4, that is, device-level package and system-level package.

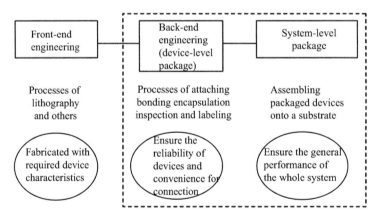

Figure 1.4   Three phases of electronics fabrication

Front-end engineering begins with the wafer. After the repetitious processes of lithography, oxidization, deposition, diffusion, etching, and passivation, integrated circuits or components are fabricated with required device characteristics.

Back-end engineering begins with the chips diced from the wafer. The chips go through the processes of attaching, bonding, encapsulation, inspection, and labeling to finally turn into the packages as components or parts, to ensure the reliability of devices and be convenient for connection with exterior circuits and circuit boards.

System level packaging in electronic engineering, including assembly technology and substrate technology, refers to assembling packaged devices onto a substrate, making good electrical connection between the devices and the substrate to form a complete system of electronic equipment so as to ensure the general performance of the whole system. In some high-density packaging engineering, bare chips can be attached directly to substrate as well.

1.5   History of Microelectronics Packaging

As shown in Figure 1.5, the history of microelectronics packaging technology can be divided into three stages.

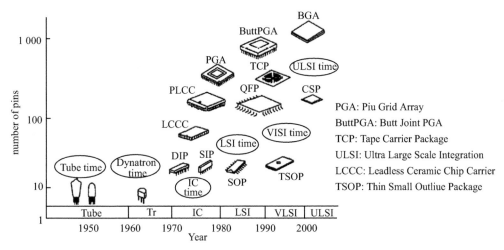

Figure 1.5   Types and feature of microelectronics packaging over time

First Stage: Prior to 1970s, mainly in-line package. Include initial metal cylinder package, the subsequent ceramic dual in-line package, ceramic-glass dual in-line package, and plastic dual in-line package (DIP), which were the mainstream products, as a result of excellent performance, low cost, and suitability for mass production.

Second stage: After 1970s, chiefly the quad package for surface mount. Surface mount technology (SMT), which was a revolution in the domain of electronic packaging, has been booming since then. Some packaging types that adapted well to SMT, such as plastic leaded chip carrier (PLCC), plastic quad flat package (PQFP), plastic small outline package (PSOP, plastic SOP), and quad leadless package, emerged. PQFP was the leading type of this period for its advantages of high density packaging, fine lead pitch, low cost, and adaptation to surface mount.

Third stage: Since 1990s, mainly area array packages. At the beginning of 1990s, integrated circuit (IC) evolved into the phase of very large scale integration (VLSI). Requiring higher density and higher speed, the IC package, therefore, developed from quad leaded to area array types. Ball grid array (BGA) was invented and became the mainstream products very soon. Since then, a variety of chip scale packages (CSP) with smaller size was developed. Also at the same period, multichip module (MCM) grew vigorously, which was also considered a revolution in electronic packaging. MCMs are classified by substrate: MCM with a ceramic substrate (MCM-C), MCM with a dielectric layer (MCM-D), MCM with a laminated circuit board (MCM-L), and MCM with a C/D (MCM-C/D). Three-dimensional package and system in package (SIP) were also developed rapidly as required by circuit density and function.

To sum up, packaging technology developed from dual in-line leads, quad in-line leads, to grid array interconnections in which packaging density was growing, and currently is developing from the two-dimensional to the three-dimensional packaging phase.

1.6   Status and Function of Microsystems Packaging Technology[1,5−8]

The importance of microsystems packaging technology is reflected in multiple levels and aspects. First, microsystems packaging technology is an important part of microsystems technology, where these technologies are closely associated with each other. For example, in microelectronics, design, fabrication, and packaging have already become three equally important parts. Excellent designers stand in more and more need of systematic concepts and knowledge of packaging technology for each of their designs. Advanced IC fabrication flow will not start until the packaging scheme is settled. Because packaging is crucial to performance, size, cost, and reliability of products, packaging technology is often the main bottleneck of device and system products.

It is well known that microsystems packaging is the fundamental technology of the information industry. Although they function differently in different products, the following six issues are common:

1. Possessing a Great Portion of Market

At the beginning of this century, the related market of microsystems packaging technology had already reached up to US$ 100 billon, accounting for 8%−9% of US$ 1.2 trillion of information industry. The main products include microelectronic circuit and device packages, packaging materials, passive components, printed circuit boards and ceramic substrates, thermal management products, cables and optic cables, flat panel displays (FPD), optoelectronics packaging, RF packaging, MEMS packaging, printing equipment, assembly equipment; and so on. About 60 billion IC and devices must be packaged and systematically assembled on board per annum.

2. Determining the Performance of Products

The performance of microsystems products is determined not only by the technical competence of the core chip but also by the whole technical competence of the system. Cannikins Law functions well in microsystems products, as packaging becoming the shortest block. For example, the clock frequency of central processing unit (CPU) chip has reached up to several GHz, which enables ultrahigh-speed signal processing, while interconnection, board; and package-related parts only allow MHz signal transmission. Therefore, realization of microsystems, which is capable of signal processing to the GHz range, relies on the development of related packaging technologies.

3. Determining the Size of Products

The feature size of semiconductor chips is moving from deep-submicron to nanometer, and the size of most chips is within a square millimeter. The size of a system is mainly determined by the IC packages, packaged modules, subsystems, and passive components, including inductance, capacitance, switch and relay, etc. The improvement of the functional density of a system totally depends upon advanced packaging technology.

4. Determining the Reliability and Life Span of Products

With the improvement of design technology, the failure rate of numerous microsystems chips has already dropped to several parts per million (ppm). Weaknesses in stress change, interconnection, and substrate interconnection, within microsystems, which are much more serious than that of chips, has been the main cause of device and system failure.

5. Determining the Cost of Products

With the rapid development of semiconductor technology, the semiconductor manufacturing technology with bigger wafers and finer lines greatly reduces the cost of chips of various kinds. The average cost of each chip has fallen below US\$ 5.0 in mass production. Millions of one dollar products have become the basic objective of market competition. Cost of packaging has reached more than 50% of that of a microsystems product, and up to 60%−90% of that of MEMS products. Costs of various packaged components and system level packages, therefore, have been regarded as the key determinant of products' market competitive power. In addition, standardized packaging also allows these components to be applied in mass engineering.

6. Determining the Technical Competence of Modern Industry

Microsystems products have reached every field of the national economy. Their mass usage and improvement of technical competence have become important aspects of technical achievement in modern industrial society: transportation, communication, office automation, home electrical apparatus, medical treatment, aviation, and avionics, and the defense industry.

1.7   Technical Challenges in Microsystems Packaging

From the systems and applications perspective the technical challenges in microsystems packaging can be divided into two major parts: low cost and high performance.

On one hand, fabrication technology of microelectronics chips, following Moore's law, is developing rapidly, currently having realized industrialized mass production of 12 inch and 0.13 µm to 45 nm. The nanometer-scale manufacturing age has been approaching, and the cost of producing a single chip is falling. Thus, the proportion of packaging expense to product cost is growing year by year. On the other hand, packaging becomes more and more of an obstacle in improving performance of microsystems to meet the demand for higher and higher performance and functional density. Tables 1.1 and 1.2 show the development of IC technology and specific demand predictions for high-performance IC packaging, respectively. Figure 1.6 illustrates the bottleneck effect of microelectronics packaging on system performance in advanced digital microsystems, which will grow with time.

As shown in Figure 1.7, microsystems packaging technology is associated with the widest technology areas. It is the most distinct in interdisciplinary characteristics and the most challenging technology in all engineering industrial technologies. Compared with the mature base of the design and manufacturing industry, the development of microsystems packaging lags behind, and thus it is hard to meet the developing needs of the industry.

It is not easy to systematically present so much professional knowledge in the limited pages of this book. Coauthored by more than 50 internationally distinguished experts in the electronics packaging industry and led by Professor Rao R. Tummala, the first monograph on microsystems packaging technology in the world was published in 2001. Within a volume of 967 pages, it presented a comprehensive discussion of major technical achievements in microsystems packaging technology. The intention of the current book is to provide students and researchers in fields related to microsystems technology, such as microelectronics and optoelectronics, with some fundamental knowledge in terms of system and applications.

Table 1.1   Technical development course and demand for packaging terminal

Year	1992	1995	1998	2001	2004	2007
Feature size µm	0.5	0.35	0.25	0.18	0.12	0.07
Gates in one chip	300 k	800 k	2 M	5 M	10 M	100 M
Number of transister cm^2	0.01B	0.04B	0.1B	0.22B	0.6B	2.1B
Chip size mm^2(logic)	250	400	600	800	1000	1250
Max. power W	10	15	30	40	40–120	40–200
Voltage V	5	3.3	2.2	2.2	1.5	1.5
I/O pins	500	750	1500	2000	3500	5000

Table 1.2   Demand for high-performance IC packaging

Year	2002	2005	2008	2011	2014
cost (cents/pin)	2.66	2.28	1.95	1.68	1.44
Power W	129	160	170	174	183
I/O pins	2248	3158	4437	6234	8758
Frequency MHz	800	1000	1250	1500	1800

From the technology chain of information system, this book focuses on microelectronics packaging and integration while incorporating MEMS packaging technology, RF packaging technology, and optoelectronics packaging technology. The contents may thus be grouped into five major parts:

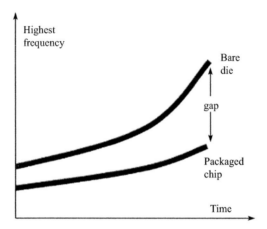

Figure 1.6   Effect of microelectronics packaging on microelectronics digital system

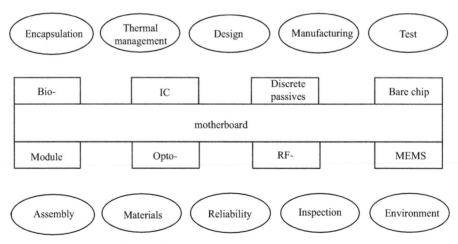

Figure 1.7   Related packaging technologies of typical microsystems

Chapter 2 discusses basic technology of microsystems packaging design, including electrical and thermal designs, mechanical design, microfluidics design, and multiple-domain design.

Chapter 3 includes basic manufacturing technology of microsystems packaging, comprising substrate technology and interconnect technology.

Chapters 5, 6 focus on device level package and integration and MEMS packaging technology.

Chapters 7 and 8 discuss basic system level package and integration technology, including module assembly, the advanced microsystems packaging technologies such as SOC, SIP, RF packaging technology, and optoelectronics packaging technology.

Chapters 1, 9 and 10 present basic concepts, testing, and reliability fundamentals, and general development trends. Readers who seek in-depth understanding of specific technology and techniques of microsystems packaging may refer to other related publications.

Questions

(1) What are microsystems? What aspects are included in the fundamentals of microsystems packaging and microelectronics packaging?

(2) What are the functions of microsystems packaging? What are the main challenges?

(3) Discuss the relationship between microelectronics packaging and microsystems packaging.

References

[1] R. R. Tummala. Fundamentals of Microsystems Packaging. New York: McGraw-Hill, 2001.

[2] Minbo Tian. Electronic Package Engineering. Beijing: Tsinghua University Press, 2003.

[3] Congming Yang. The Packaging Process of Metal Microcap under Room Temperature Status and Its Shear Stress-Strain Relationship Analysis. Kaohsiung: National Sun Yat-sen University, 2004.

[4] Mei Li. Development and the prospect of microelectronics packaging technology. Semiconductor Technology, 5 (2000): 1–3.

[5] Robert Markunas. MEMS: Mainstream integration. Electronic Engineering and Product World, 4 (2004): 31–34.

[6] Sulong Tu, Xinbo Huang. Technologies and applications of MEMS. Journal of Qingdao Technological University, 3 (2003): 73–78.

[7] Zhiyi Tong. MEMS micromachining technology and tools. Equipment for Electronic Products Manufacturing, 1 (2004): 5–11.

[8] R. R. Tummala. Packaging: Past, present and future. 6th International Conference on Electronic Packaging Technology (2005).

CHAPTER 2

Design Technique for Microsystems Packaging and Integration

Design is one of the key technologies in microsystems packaging, similar to manufacturing, test, and failure analysis. In addition to having an idea on the packaging scheme to be used, an excellent packaging designer must also have versatile knowledge, such as chip design and fabrication, system requirements and assembly, reliability and application requirements, and cost control, among other factors. This chapter depicts the basic design techniques, including electrical design, thermal management design, mechanical design, high-frequency (HF) design, microfluidicdesign, and multidiscipinary design.

2.1 Electrical Design[1−9]

2.1.1 Overview of the Electrical Design in Packaging

A successful electrical design for microsystems packaging means that electrical performances, including electrical signals and power flow, can satisfy the whole system requirement after packaging. The deliverables of electrical design are the interconnection layout, material selection, and geometric dimensions of different components. In some cases, embedded passive components and embedded optical waveguides should also be taken into consideration in electrical design. In principle, electrical functions of packaging can be divided into two major parts: signal distribution and power distribution.

2.1.2 Fundamentals and Flowchart of Electrical Design

Fundamentals of the electrical design include electrical principles, circuit signal processing, electromagnetics, etc.

1. Power Distribution

(1) Electrical resistance. Current will dissipate power in the form of heat when traveling through a resistance, and thereby, the power amplitude will vary accordingly. Therefore, it is all-important to minimize the electrical resistance of wires and other conducting paths. These problems are usually analyzed using Ohm's law and Kirchhoff's voltage law.

(2) Simultaneous switch noise (SSN). Since the power supply voltage always fails to make a real-time response, the local direct current (DC) power's voltage at some points in the system will drop transitorily, and thereby cause the signal fluctuation. This fluctuation is called simultaneous switch noise (SSN). To suppress such an electrical noise, a decoupling capacitor should be designed properly.

2. Signal Distribution and Wiring

(1) Skin effect. A DC current travels through the conductor's cross-section uniformly, while alternating current (AC) at high frequencies tends to travel near the surface of the conductor. AC resistance of a conductor is much larger than DC resistance. This is called skin effect.

(2) Parasitic capacitance, parasitic inductance, and noise. Parasitic capacitance is the

inherent capacitance existing between any resistance pair in a circuit system. Similarly, parasitic inductance is engendered in any conductive current structure, and noises are nonideal effects of all kinds occurring in the system, which will create distortion of wave shape and signal amplitude. Parasitic capacitance and inductance are two important causes of the noises, and they should be reduced by the electrical design.

(3) Time delay. Either combination of resistance and capacitance (RC) or that of resistance and inductance will cause time delay in a system, and this time delay will result in a nonreal time signal transport in the circuit. Because most interconnects contain resistances and a considerable amount of capacitance, the RC delay, therefore, becomes a restrictive factor for the system's speed. Consequently, RC delay should be given serious consideration in the electrical design. Furthermore, the time delay caused by the coexistence of resistance and inductance will not only affect the real-time response of a power-on-chip to the changed orders sent by the circuit, but also generate SSN in the circuit.

(4) Transmission line. Usually interconnects on-chip can be simulated as a RC circuit, while those in the printed circuits and packaged bodies have to be simulated as transmission line due to size issues. Any pair of conductors that transmit signals or power can be regarded as transmission line. Problems concerning transmission lines are generally analyzed and solved based on electromagnetics.

The key parameter of the transmission line's performance is its characteristic impedance, which designates the ratio of voltage to current of an electromagnetic wave traveling along a given direction. Characteristic impedance of a transmission line is a function of transmission line material and its geometric size. Discontinuity of the characteristic impedance will lead to a reflection in a transmission line locally, resulting in an additional noise being generated. Therefore, design should be made on the end of a transmission line so as to make the load impedance match with characteristic impedance of the transmission line.

(5) Crosstalk phenomenon. Crosstalk noise results from the induction on a specific line, which originates from signals on another adjacent line, though there is no physical contact between the two. Crosstalk is a complicated phenomenon, which is usually caused by parasitic capacitance and inductance. Transmission line models can be used to predict and reduce reflection and crosstalk noise.

3. Electrical Design Flowchart of a Package

An electrical design of a package usually includes the following steps:

(1) Determine electrical parameters of a package according to system requirements and chips specifications, including chip interconnections, output characteristic of circuits, communication speed, power consumption distribution of circuits and chips, etc.

(2) Determine physical parameters of package according to design rules and characteristics of the materials used, such as minimum line width, interconnect spacing, thickness of transmission line and dielectric layer, dielectric constant, and electrical conductivity.

(3) Determine the parameter extraction tool and use it to extract the electrical parameters and parasitic parameters like R, L, C, G, etc. The commonly used tools include Maxwell, EM, HFSS (Ansoft Corporation), etc.

(4) Use simulation software like simulation pragram with integrated circuit emphasis (SPICE) and parameters mentioned above to construct a circuit interconnecting model.

(5) Implement simulation of signal distribution and power distribution in package to obtain detailed distribution situation of signal reflection, crosstalk, signal attenuation, DC depreciation, and power supply noise.

(6) The wiring rules, delay equation, and noise scheme are combined with computer aided design (CAD) tools to determine the physical layout and process rules of the packaging process.

To optimize the packaging structure, the aforementioned simulation should be conducted individually according to the impedance distribution and noise target. Further simulation-based optimization should cover fore and aftparts interconnecting package signals and power supply. Consideration should also be given to static performance, actual temperature variation, and details at every node in the package.

2.1.3   Analysis of a Package's Electrical Performance

A typical electrical function of a package is shown in Figure 2.1. Packaging should provide a signal path between chips, power and ground wire interconnecting of chips, and interconnection between chips and passive components. As shown by the dashed line in Figure 2.1, the signal path is composed of soldering pads at the interface between chips and package, transmission lines on substrate, and via holes providing vertical interconnection. Through this signal path, chips can execute exchange of data, address, time control, and other electric signals. Steady and accurate power and ground signal will guarantee the normal operations of the active components. Therefore, the parasitic inductance induced by power distribution and soldering pads has to be minimized to ensure that they do not contribute to the performance degradation of power-supply voltage. This can be achieved by using a planar power cord in a package, which can provide a larger capacitance and a smaller inductance at the same time.

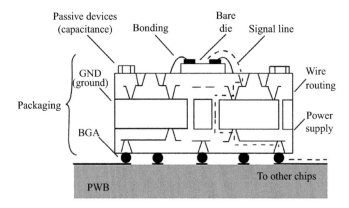

Figure 2.1   Schematic illustration of electrical function of a package, PWB, printed wiring board; BGA, ball grid array

The first level package is often used as the converter between chip interconnect and circuit board interconnection, that is, the spatial converter between the subtle interconnection structure of chips of submicron scale and the lead structure of the circuit board on a scale of tens of microns or above. Chips may have hundreds of I/O pins, which need to be connected to soldering pads on circuit boards. Interconnections inside a package are, accordingly, as intricate. In order to contain these interconnections, multiple signal wiring layers have to be used.

Since the geometrical shape of interconnections has little effect when they work in low frequency, the wiring of the signals and power supply can be easily achieved. However, at high frequencies above hundreds of MHz, interconnections are actually much longer than the energy path along them, and their behavior characteristics are determined by material characteristics and the electromagnetic fields, which constitutes signals. Some effects, like transmission delay, characteristic impedance, and parasitic reactance related to interconnection structure, are responsible for the signal characters. Therefore, both the distortion rate of a signal and the time it takes for the signal traveling from the starting point to the terminal are the functions of the interconnection's parameters. Moreover, the power supply and the ground path need to be taken care of.

Systematic electrical specifications of a package include the following parameters: delay, distortion, load, impedance, reflection, crosstalk, undulation of power/ground, etc. Some circuit simulation methods based on the circuit model of the packaging structure are used to evaluate the electrical performance of a package. The circuit model for physical structure can be obtained by using electromagnetic simulations, analytic equations, or the HF measurement method.

1. Modeling

Nowadays, there are many commonly used circuit analysis tools available for circuit modeling. One of them, SPICE contains all the models of passive components and most active components and can be used to simulate the time and frequency domains.

First, the coupling parameters of the function structures are extracted, such as signal lines, solder balls, and through-holes in package, whose resistance and reactance will lead to an increment of parasitic effects on signal lines and degradation of the signal quality. Figure 2.2 shows the electrical equivalent of some parasitic effects. Second, passive components installed on the package and circuit board can be modeled by some equivalent circuit, for instance, a decoupling capacitor can be expressed by a resistance, inductauce, capacitance (RLC) serial circuit.

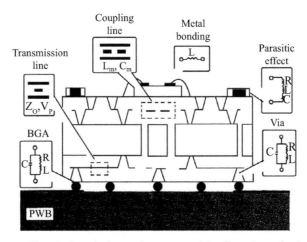

Figure 2.2   Electrical equivalents of some parasitic effects in packaging design

The major challenge in electrical design is to deal with various nonideal effects that impact electrical signal transmission. Typical nonideal effects always result from non zero transmission delays of actual electrical signals and parasitic reactance of all kinds occurring in every circuit. Electrical design is required to make a tradeoff between noise reduction and complexity of mask layout so as to work out a package satisfying the system performance requirements.

2. Analysis of Signal Distribution

Signal transmission refers to data or commands being transmitted from one point to another in a system. Starting from the driver circuit of a certain chip, a signal travels through interconnects to the receiving circuit in either the same chip or another chip. Interconnects can be implemented on chips, package, or printed wiring board (PWB). In design, consideration should also be given to the following issues.

(1) Capacitive delay of interconnection. Interconnects and metal conducting wire of the ground circuit can construct some capacitances, and an interconnect itself can be regarded as a charging or discharging capacitor. Stray capacitance will also be introduced as a result of the physical proximity effect of other interconnects, soldering pads connecting the integrated circuit with its package, and bends of wiring. Since the geometric size of the interconnects in the package is considerably bigger than those in the microchip, the capacitive effect of this kind is large enough to merit serious consideration. The capacitive effect of the interconnection can be analyzed in Figure 2.3, where C, C_g, and R_{on} stand for capacitance between the interconnects the and ground, equivalent capacitance of the receiver ingress/threshold, and equivalent resistance between the circuit and the capacitance in the chip, respectively. V_{dd} and V_{load} are the signal input voltage and the load response voltage. The latter can be calculated through Equation 2.1 as follows.

$$V_{load}(t) = V_{dd}\left(1 - e^{-t/\tau}\right) u(t). \tag{2.1}$$

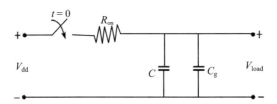

Figure 2.3   Capacitive effect of interconnection

Example: When a package has the following structural condition: $V_{dd} = 5$ V, $R_{on} = 50\ \Omega$, and $C + C_g = 10$ pF, the characteristic time of the interconnection and the capacitance can be calculated as $\tau = R(C + C_g) = 0.5$ ns. The delay related to capacitor charging makes the voltage on-load reach 50% of its stable value until $t = 0.35$ ns and 90% until $t = 1.15$ ns. Generally serious consideration should be given to the time delay in the communication between circuits caused by the capacitor charging to 90% of its stable value.

(2) Delay equation. The digital circuit operates based on threshold logic, which means that the circuit's state will change as the input reaches a certain level. In most circuits, the threshold level is equal to 50% of the logic swing. When output reaches 50% of input value, the delay may be expressed as

$$T_{50\%} = 0.69 R_{on}(C + C_g). \tag{2.2}$$

This is termed the delay equation, which is derived by assuming interconnection is a lumped capacitance. The delay equation is usually used to determine the timing information of the circuit. The actual delay equation is much more complicated than Equation 2.2, since all parasitic capacitance delays have to be taken into consideration in a practical application.

3. Power Distribution

There are two important issues to discuss in power distribution: IR voltage drop and inductive effect. A voltage drop occurs when DC current travels through resistance R. The resistance of package metal includes through-hole resistance, interconnection resistance, and planar lead resistance. Owing to the occurrence of the voltage drop, the power-supply voltage inside the chip circuit might be position dependent. When this distorted input becomes serious, the circuit will switch incorrectly.

Inductive effect is another important issue in the power distribution design. All physical circuits bear a certain inductance, which is in series with power supply, since power supply transmits current to each part of the chip through a package. Generally, the inductance related to the power-supply circuit and the ground circuit cannot be neglected. For instance, Figure 2.4 shows power-supply voltage V_{supply} driving the P-channel Metal Oxide Semiconductor (PMOS) transistor load (equivalence to the chip), where L_1 and L_2 denote inductance of the power supply and that of the ground wire in the package, respectively. The switch refers to the transistor with a zero on resistance, and it also drives the load with a resistance of R Ohm. This circuit describes how package inductance provides power for the chips.

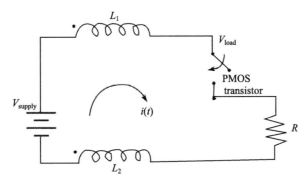

Figure 2.4   Inductive effects in a typical package

Load voltage applied to the chip is

$$V_{load}(t) = V_{dd}\left(1 - e^{-t/\tau}\right)u(t),\tag{2.3}$$

and time constant, τ, is given by

$$\tau = (L_1 + L_2)/R.\tag{2.4}$$

To illustrate the package inductance effect, assuming $V_{supply} = 5$ V, $R = 50\ \Omega$, and $L_1 = L_2 = 1$ nH, the time constant is 40 ps. The inductive effect of the power distribution network in the package will induce a considerable delay on high-speed circuits.

4. Power-Supply Noise

The noise produced in the power distribution network is likely to decrease or increase power-supply voltage when circuits are switching on or off simultaneously. The decrease makes the voltage between V_{dd} and GND on the chip less than the power-supply voltage, while the increase has the opposite effect. Both effects will degrade the performance of the system. Besides, the inductance in the package and the capacitance in the chip may result in resonance, thus evoking oscillation of power supply.

Because all of these are non random events, the output-driver circuit can cause much more noise than the internal circuit. Moreover, the existence of inductance in interconnection requires more current to charge up. For a high-speed system, the power-supply distribution network of the on-chip circuit needs to be isolated from that of the off-chip circuit.

Besides, with the increase of circuit speed, current transfer becomes faster and faster, and the on-chip current may experience considerable changes over short intervals, and consequently dI/dt can be very big, thus causing oscillation of power supply. With the trend of reducing voltage and signal level gradually in chip design, power-supply noise also needs to be depressed.

The return current path also has effects on the noise in the output-driver circuit. For a computer system running in high frequency (HF) the signal wave transmitted along a plane reflects on the edge of the package so as to evoke oscillation within the structure, which fuels the noise in the package and cannot be depicted by the model of effective inductance (L_{eff}). A distributed model or a transmission line model is required to estimate the power-supply noise in the modern high-speed system. Presently, modeling and simulation of SSN is an applicable but challenging research subject in high-performance packaging design. Since thousands of through-holes and interconnections are contained in a package, it is very difficult to measure the current path one by one, and therefore software is usually used to extract parameters such as effective inductance, L_{eff}.

It should be noted that in the second level package, the motherboard supplies chips with voltage and current through the first level package as well as the metal layer in the PWB. The package and the metal layer in the PWB distributing network for power will cause inductance to increase, and, accordingly, the current path from the power supply will be extended or probably bypassed.

5. Electromagnetic Interference (EMI)

In electronics packaging, electromagnetic interference mainly includes electromagnetic radiation in free space and direct noise transmission through power/ground wires and signal wires. When the clock frequency of digital circuits and systems becomes higher and higher, HF noise and resonance of the basic clock signal will occur in components, package, and system. This will bring about remarkable radiation and makes an EMI-free design much more difficult.

Generally, EMI sources exist in the PWB, cables, the chassis of wireless frequency devices, and high-speed digital systems. However, when frequency increases, the devices and the package must also be regarded as EMI sources, which include reflection along signal wires and transient change on the power supply/ground wire. The most notable radiation structure is large conducting wire, such as planar interconnects and cables between circuit boards, which, as shown in Figure 2.5, can act as antennas to radiate energy. The device package is usually much smaller than the circuit board or the cable in size, and therefore, it is not a radiation source unless a package is much bigger than normal, with high noise frequency. To sum up, the basic points for packaging designers to consider in the EMI issue are the frequency capacity and the amplitude of noise current.

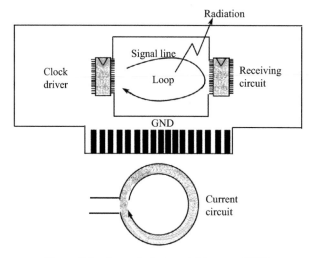

Figure 2.5   An equivalent aerial loop on PWB

How to control EMI is a complicated issue involving many aspects of design and requiring concerted efforts of different designers in chips, package and system. The following should be noted in the electrical design:

(1) Improper layout of the decoupling capacitor, a big current loop, and ground voltage fluctuation will lead to a detrimental radiation reflection.

(2) For signal wire exposed on the surface of a PWB, such as microstrip structure, impedance mismatch induced reflection in signal wires must be avoided.

(3) Discontinuity of capacitance along transmission lines will result in an undershoot of the signal waveforms, while discontinuity of inductance will lead to an overshoot of signal waveforms. Therefore, when the transition time of a digital signal becomes shortened, parasitic effects in the package and the resistance used to match transmission line must be closely controlled.

(4) In the distribution of package pins, incorrect choice of ground wire pins is also a source of EMI. To reduce the radiation emission to a minimum, each signal wire and ground return path must be restricted to a very small size. It is especially important that the area of ground loop on the PWB should be close to the signal wire so as to form a current path with a smaller loop.

(5) As a result of SSN, inductance in the ground wire network of a package may cause voltage fluctuation. Potential differences between a certain nonideal grounding point and another grounding point might produce a large high-frequency grounding current. If a return loop is formed, grounding current can then be regarded as an antenna loop; thus EMI appears.

(6) On a circuit board, if the sheath of the cable is connected to a grounding point with noise, the cable sheath then becomes a half-dipole antenna, which makes this issue much worse.

2.2   Thermal Management Design[1,2]

2.2.1   Overview of Thermal Management

Thermal management in this book means using proper packaging materials and packaging methods to realize proper heat transfer conditions, and thereby make components, devices, and systems work in an appropriate temperature environment.

There are two main heat sources in a package. The first one is the heat transformed from the electric energy of a current inside the package; leads, resistances, and polysilicon will generate heat when a current is applied. Some key components with a high functional density, such as a CPU, have a power consumption larger than 100 W. The second is the heat generated by the friction of moving parts, such as micromirror array. For a package with constant heat generation inside, the temperature will increase at a constant rate until the device stops working or loses its physical properties, if no cooling methods are adopted. Thermal management tries to export the heat and maintain the temperature of the package within a permitted range. The main methods of thermal management include conduction, convection, radiation, and phase-change process. In the thermal management of microsystems packaging, careful consideration should be given to the system application requirements and heat generation mechanisms inside chips. Moreover, the right combination of packaging materials and heat transfer mechanisms should be designed optimally. The following content will focus on the thermal management of the microsystems packaging. The principal goal of the thermal management of electrical products is to avoid instantaneous failure or full failure of the chips' electrical functions and the package.

2.2.2   Importance of Thermal Management

Thermal design is one of the most important facets of the packaging design.

First, good thermal design can prevent thermal damage to the system. Without thermal management, the system performance will decline, temperature will increase rapidly, stress will rise, the structure will be destroyed, and melting, evaporation, or burning can occur. In practical applications, excessive temperature is a major failure factor for many microsystems.

Second, the progress of thermal design is promoted enormously by chip technology. The great increase of the circuit density requires research on novel and high-efficiency thermal management technology. Dissipation power of the chip has risen from 100–300 mW of small scale integrated circuit (SSIC), 1–5 W of very large scale integrated circuit (VLSI), to tens of Watts currently; the power consumption of Microprocessor unit MPU has ascended from 15–30 W in the 1990s, 50 W in 2000, to 100 W currently. With integrated circuit growing in scale, a single microelectronics chip contains tens of million transistors and thousands of pins, and multichips are applied in a single package. Therefore, it is a severe challenge to ensure that products work correctly within their lifetime.

Similarly, research on thermal damage mechanisms of specific devices and systems would increase the likelihood of adopting reasonable cooling methods and heat transfer models to establish thermal control strategies early in the design process. Although different in size, power consumption, and temperature sensitivity, microelectronics components have similar aims and technologies in terms of thermal management, and thereby demand as much consideration. For instance, working for a long time in a high temperature environment, many microcircuits suffer from loss of circuit speed, timing failure, and even failure. For example, mechanical creep, parasitic chemical reaction, and adulterant diffusion of bonding materials will all accelerate failure. The related failure mode indicates that the failure rate of electrical components has an approximately exponential relationship with the working temperature.

2.2.3   Basics of Thermal Management

1. Conduction

(1) Equation of heat conduction.

In the solid, static liquid, or gas, thermal energy transfers from the high-temperature region to the low-temperature region, and this process is called conduction. Conduction is caused by direct intermolecular energy exchange. Most conduction problems in the package can be expressed by the following one-dimensional Fourier equation,

$$q = -kA\frac{\mathrm{d}T}{\mathrm{d}x}, \tag{2.5}$$

where q is the heat flux in watts, k is the thermal conductivity in W/(m·K), A is the cross-section area where the heat flow travels through (m^2), and dT/dx is the temperature gradient in the direction of heat flow, whose unit is K/m.

Thermal conductivity, k, is a physical property to describe the ability of heat transfer in a material. Typical electronics packaging materials have totally different conductivities: for air, k is 0.024 W/(m·K); silicon, 150 W/(m·K); and synthetic diamond, 2000 W/(m·K). Compared with air, the thermal conductivity of polymer is more than 10 times greater, water 25 times, ceramics 100 times, metals 10,000 times, and diamond up to 83,300 times. Most IC packages and printed circuit boards (PCB) are composed of epoxy resin and polymer with low thermal conductivities. The only way to transfer heat from the package to outsides is to attach heat-sink through-holes and connect them to materials with high thermal conductivities like metallic slices.

(2) Lumped heat capacity.

In electronics packaging, the package temperature will increase with heat dissipation inside. If the temperature inside a solid increases slowly enough, the whole solid can be treated as having the same temperature. This relationship is often called lumped heat capacity, and it is generally used when the solid has a high thermal conductivity. For solids with high thermal conductivities and with an internal heat source, without any external cooling, temperatures will increase at a constant rate, which can be calculated as

$$\frac{dT}{dt} = \frac{q}{mC_{\mathrm{p}}}, \tag{2.6}$$

where q is the internal source in watts, m is the mass of the solid in kg, and C_{p} is the specific heat capacity of the solid in J/(kg·K), respectively.

(3) Thermal resistance.

In thermal design, conduction is often compared to current flowing along a conductor. When comparing heat flow q and temperature drop ΔT to current I and voltage drop ΔV, a thermal Ohm's law can then be obtained and thermal resistance can be defined as below:

$$R_{\mathrm{th}} = \frac{\Delta T}{q}. \tag{2.7}$$

Strictly speaking, such comparisons can also be extended to other heat transfer models even though it solely serves as conduction.[10] According to Equation 2.6, R_{th} is in inverse proportion to the mass and the specific heat capacity of the solid, which is often determined via experimental methods.

Comparing Equation 2.5 with Equation 2.7, thermal conduction resistance can be derived as

$$R_{\mathrm{th}} = \frac{L}{kA}, \tag{2.8}$$

where L is the length of a single material. Figure 2.6 shows the thermal conduction resistances of typical packaging materials.

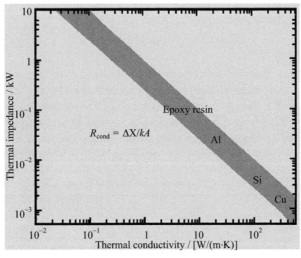

Figure 2.6   Conductive thermal resistances of typical packaging materials

For instance, a copper plate with an area of 100 mm^2 and a thickness of 25 μm has a thermal conduction resistance of 0.00060°C/W.

In the thermal design of a practical component, the overall thermal resistance of composites in the package can also be calculated by using the same method as electrical resistance calculation of serial and parallel circuits. The calculation method of serial thermal resistance is applicable to structures where the normal direction of the package plate is parallel to the thermal conduction direction, while that of parallel thermal resistance is applicable to structures where the normal direction is vertical to the thermal conduction direction. Similarly, more complicated systems can be discussed by a thermal resistance network.

2. Convection

In a package, heat transfer from the solid (package) to the fluid (space or cooling media) should always be given careful consideration. This is convection, and heat dissipation through a fan, as shown in Figure 2.7, is a typical form of convection. Convection has two basic heat transfer mechanisms: one is heat exchange by quasistatic molecules surrounding the solid's surface, which is similar to thermal conduction; another one is heat transfer by fluid motion as a whole on the solid surface. Convection heat transfer can be described by a linear relationship between heat flux and temperature difference, which is known as Newton's law of cooling.

$$q = hA(T_s - T_f), \tag{2.9}$$

where h is the heat transfer coefficient, A is the area of wet surface, and T_s and T_f are the surface and the ambient fluid temperatures, respectively.

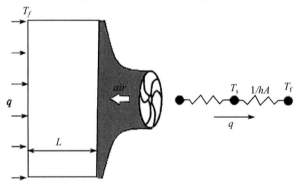

Figure 2.7   Convective heat transfer

The difference in heat transfer between fluids circulating with high and low velocity and convection performance of different fluids can be characterized by the convective heat transfer coefficient, h. Table 2.1 shows the variation range of convective heat transfer coefficients of typical fluids.[1,2,10] There are a lot of theoretical formulae and empirical relations in the heat transfer literature, and they can be used to determine convection heat transfer coefficients of a given fluid on different grooved surfaces or various geometrical bodies. [10,11]

Table 2.1   Typical values of convective heat transfer coefficient (HTE)

Mode of convection	HTE/$[\text{W}/(\text{m}^2 \cdot \text{K})]$
Natural gas convection	5−15
Natural liquid convection	50−100
Forced gas convection	15−2500
Forced liquid convection	100−2000
Boiling liquid or condensed vapor	2000−25000

3. Thermal Radiation

To absorb and emit thermal energy through electromagnetic waves and photons is another kind of heat transfer mode called radiation heat transfer or thermal radiation. Unlike conduction and convection, thermal radiation can occur in any medium, even in a vacuum, and the radiation flux is proportional to the temperature difference between the heat source and the heat dissipating body to the power of four,

$$Q = \varepsilon \sigma A \left(T_1^4 - T_2^4 \right) F_{12}, \tag{2.10}$$

where ε is the thermal emissivity; σ is the Stefan-Boltzmann constant, whose value is $5.67 \times 10^{-8} \mathrm{W}/(\mathrm{m}^2 \cdot \mathrm{K}^4)$; and F_{12} is the angle factor between surfaces 1 and 2. For the case where two surfaces that have a high absorptivity and a high emissivity are very close to each other, or that where heat radiates from one surface with a high emissivity to another has a much larger surface area that nearly enclosing the former fully and a high absorptivity, F_{12} is approximately 1. The relationships between the angle factors and the complex geometrical shapes, as well as the various surface conditions, are available in references.[11−13]

2.2.4   Electronic Cooling Methods

Different packaging applications need different cooling methods. In addition to how much heat should be transferred, thermal management design will consider the system application environment and the category, volume, and cost of cooling techniques. Thermal management methods are grouped with passive cooling and active cooling. In the applications with passive cooling techniques, highly thermal conductive glue is adopted in the general IC chip package. In the device-level package, grease, phase change materials (PCMs), heat sinks, heat pipes, and the like are used for cooling. In the active cooling technique applications, widely adopted measures include fans, liquid medium convection, thermal to electric converters, condensers, air conditioning technology, etc. Some basic elements of the cooling methods are presented briefly below.

1. Heat Sinks

Thermal resistance can be reduced by increasing either the heat transfer coefficient or the heat transfer area. In thermal design, this is realized by increasing the surface area, i.e., adding fins. Figure 2.8 shows a heat sink, which is widely used in electrical systems. Of course, in practical applications, there are groups of such heat-sink fins.

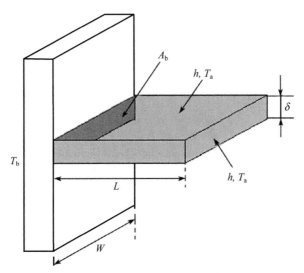

Figure 2.8   Schematic illustration of heat sink

In Figure 2.8, T_a is the base plate temperature of the heat-sink fin, T_b is the ambient temperature, and A_b, L, W, and δ are the area, length, width, and thickness of the base plate. The two heat transfer surfaces are exposed in the liquid with a convective heat transfer coefficient of h. The cross-section area, $W\delta$, is replaced by the area of the increased heat-sink fins, $2\,(WL + W\delta/2 + L\delta)$. The increment of the heat transfer area means a higher heat transfer capability of the heat-sink fins.

A precise analysis reveals that the efficiency of rectangular fins can be calculated as[14]

$$\eta = \tanh\,(mL)/(mL) = \left(\frac{\mathrm{e}^{mL} - \mathrm{e}^{-mL}}{\mathrm{e}^{mL} + \mathrm{e}^{-mL}}\right)\Big/ mL, \tag{2.11}$$

where m is

$$m = \sqrt{\frac{hP}{kA_b}},$$

where P is the perimeter of the fin. As for a single fin, the optimized heat-dissipating effect can be obtained when the efficiency is 0.63.

Cooling holes are embedded in the substrate to reduce the thermal resistance, and the best cooling performance will be obtained by arranging the cooling holes vertically to the PWB plate. The $Q_{Z,z}$ model, as shown in Figure 2.9, can be used to illustrate the case when there are a large number of cooling holes. In this case, the equivalent thermal conductivity along the direction of Z is[2]

$$k_{Z,z} = k_M a_M + k_I(1 - a_M), \tag{2.12}$$

where k_M, k_I are the thermal conductivities of the metal and the insulator, respectively, a_M is the area percentage of metal holes to the whole cross-section.

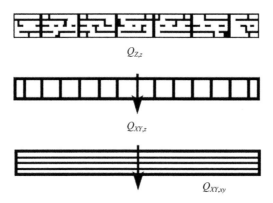

$$Q_{Z,z}$$

$$Q_{XY,z}$$

$$Q_{XY,xy}$$

Figure 2.9 Models of wiring substrates with cooling holes

For the sparsely distributed cooling holes, heat conducts mainly along the direction of XY, which is illustrated by the $Q_{XY,xy}$ as shown in Figure 2.9.

$$k_{XY,xy} = k_M t_M + k_I(1 - t_M), \tag{2.13}$$

where t_M is the thickness percentage of the metal to the PWB.

2. Heat Pipe Cooling

Heat pipe is a simple heat transfer device, which transfers heat to a point far away from the heat source through the phase change process and vapor diffusion without using any moving component. This cooling technique finds a mass of applications in which high thermal conductivity, low mass, and small size are required, such as a notebook computer.

As illustrated in Figure 2.10, a heat pipe is an adiabatic slender tube with inner structures that contains a small quantity of heat transfer medium, such as water. A typical heat pipe consists of an evaporator that absorbs heat to evaporate the liquid inside, a condenser that condenses the vapor to dissipate the heat, and an adiabatic compartment through which the medium flows. The vapor and the liquid flow inversely along the inside wall and the inner structure. This process can form a high heat transfer coefficient and can transport vapor in the evaporator with only a slight pressure difference.

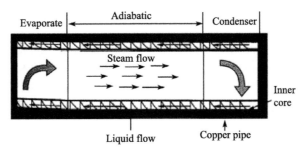

Figure 2.10   Longitudinal cross section of a heat pipe

3. Thermoelectric Cooling[12−15]

Thermoelectric cooler (TEC) is a solid-state cooling device, which works based on the Peltier effect. As shown in Figure 2.11, when a voltage difference exists at two terminals of a pair of semiconductor materials, heat absorbed into the cold terminal will transfer to the hot one, thus inducing a temperature drop in the cool terminal (package) and a temperature increase in the hot terminal (heat sink). The heat flow is in proportion to the working current applied.

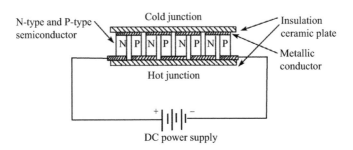

Figure 2.11   Structural diagram of a thermoelectric cooler

The main advantages of a TEC include

(1) The most notable advantage: have the capability to lower the temperature below the environmental temperature.

(2) Precise temperature control: using a closed-loop temperature control unit, a precision of ±0.1°C. can be reached.

(3) High reliability: solid-state refrigeration modules and no moving parts result in a low failure rate.

(4) Long life: over 2×10^5 hours.

(5) Operation without any noise: as opposed to the mechanical refrigeration system, there is no noise generated during its operation.

A typical TEC device is comprised of numerous thermoelectric cooling units, which are serial in terms of electrics, while parallel in terms of thermophysics. Each two thermocouples are electrically insulated by a heat-conducting ceramic sheet.

The performance of TE cooler is usually evaluated by Figure of Merit (FOM), namely,

$$\text{FOM} = \frac{\alpha_\text{s}^2}{\rho_\text{TE} k_\text{TE}}. \tag{2.14}$$

where a_s is the Seebeck coefficient, and ρ_TE is the electrical resistivity of TEC. Typical values of FOM range from 0.02 K^{-1} to 0.05 K^{-1}. Recently, the FOM of the MEMS-based TEC samples have had a magnitude increase.

Standard design rules are generally adopted in thermal management design, namely, a whole-set selection of TEC, heat-sink fins, and power supply for TEC. To increase cooling capability, TECs need to work in parallel with each other. For comparatively low temperatures in chips, TECs need to be connected in serial.

With the progress of MEMS technology, integration of miniaturized TEC and microsystem chips is close to being realized. Figure 2.12 illustrates the array elements of such a novel miniaturized TEC, whose refrigerating coefficient of performance is 0.3 and can be used in wireless communications directly.[14]

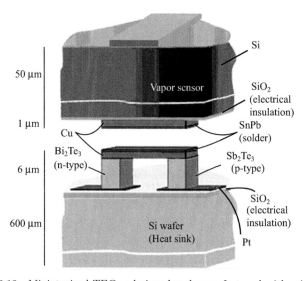

Figure 2.12   Miniaturized TEC codesigned and manufactured with microsensors

2.3 Mechanical Design[15−18]

2.3.1 Importance of Mechanical Design

In order to further reduce the package's signal delay and improve its density between circuits so as to build on the rapid development of the integrated circuits, high-precision interconnections, fine packaging structure, and high functional density should be taken into consideration. With the functional density increasing, the problems of thermal stress, thermal-mechanical failure, and mechanical performance failure become increasingly serious, and studying the mechanical performance becomes necessary. Especially with the development of multipins and fine pitches of bare chips packaging, and the microminiaturization of the multilayer surface electrode for mounting, package strength becomes increasingly important, and the issue of reliability becomes more and more prominent. At the same time, with the adoption of new technologies, materials, processes, and structures, a great change will take place in micromechanical performance, especially microcracks induced by stress, structural deformation of components, and substrates induced by temperature variation, all of which can result in increased stress problems.

2.3.2   Basic Concept in Mechanical Design

Mechanical terms include: thermal mismatch, stress, deformation, strain, Hooke's law, stress yield strength, fatigue, etc.

In microsystem packaging design, the stress change mainly comes from the following:

(1) Thermal condition's variation.

Electronics packages contain different kinds of materials, and the temperature distribution is usually not uniform. Thermal stress will occur inside the package during the manufacturing process because of mismatches in geometrical shape, thermal expansion coefficients, and material compound in the different parts. Therefore, mechanical design usually has a close relation ship with thermal management.

(2) External force.

On one hand, external forces come from manufacture, assembly, test, and usage. On the other hand, numerous force-sensitive structures in MEMS devices, such as inertial devices and mechanical sensors, can with stand different external forces.

(3) Inherence in microsystems.

A large number of MEMS devices, including MEMS switches, relays, micropumps, microvalves, microjets, micromirrors, etc. cannot work correctly unless they are driven by corresponding forces, such as electrostatic force, electromagnetic force, fluid pressure, mechanical force generated by piezoelectric materials, and thermal driving force. High temperature and pressure may occur in some local areas.

2.3.3   Mechanical Design Method

Mechanical analysis is an old but young technique. Various theories and experimental methods, solutions for specific problems, and high-efficiency computing tools cannot be presented seriatim here. In general, mechanical design methods can be divided into three categories: theoretic analysis solutions, numerical calculations, and computer-aided analysis.

1. Theoretical Solutions for Mechanical Problems

In a Cartesian coordinate system, the control equations of thermal stress and strain theory can be written as

$$\nabla^2 T = \frac{\rho C_v}{k}\frac{\partial T}{\partial t} - \frac{W}{k}, \tag{2.15}$$

$$\frac{\partial^2 u}{\partial x^2} + \frac{\partial^2 v}{\partial x \partial y} + \frac{\partial^2 w}{\partial x \partial z} + (1 - 2\nu)\nabla^2 u + \frac{X_x}{\lambda + G} = 2(1 + \nu)\alpha\frac{\partial T}{\partial x}, \tag{2.16}$$

$$\frac{\partial^2 v}{\partial y^2} + \frac{\partial^2 u}{\partial x \partial y} + \frac{\partial^2 w}{\partial y \partial z} + (1 - 2\nu)\nabla^2 v + \frac{X_y}{\lambda + G} = 2(1 + \nu)\alpha\frac{\partial T}{\partial y}, \tag{2.17}$$

$$\frac{\partial^2 w}{\partial z^2} + \frac{\partial^2 u}{\partial x \partial z} + \frac{\partial^2 v}{\partial y \partial z} + (1 - 2\nu)\nabla^2 w + \frac{X_z}{\lambda + G} = 2(1 + \nu)\alpha\frac{\partial T}{\partial z}. \tag{2.18}$$

Heat transfer and thermal stress in the electronics packaging occur in two stages: transient state (switch on/off) and steady state (in operation). In the two stages, as for thermal stress and strain theory, temperature distributions, $T(x_i, t)$, in the package are obtained from the solution of thermal conduction Equation 2.15 under certain initial and boundary conditions, while displacement components, u, v, and w, in the package are determined from the solutions of Equations 2.16, 2.17, and 2.18 under certain boundary conditions of stress and displacement, and with temperature distribution obtained above as an essential boundary condition. Thermal stress components $(\sigma_x, \sigma_y, \sigma_z, \tau_{xy}, \tau_{yz}, \tau_{zx})$ and strain components $(\varepsilon_x, \varepsilon_y, \varepsilon_z, \gamma_{xy}, \gamma_{yz}, \gamma_{zx})$ can be achieved from the solution of displacement partial differential equations.[2]

In Equations 2.15−2.18, u, v, and w are displacement components in the directions of x, y, z, respectively; X_x, Y_y, Z_z are the force components in the directions of x, y, z, respectively; α is the linear coefficient of thermal expansion; k is the heat conductivity; ν is Poisson's ratio; λ is the Lamé constant; G is the shear modulus; T is the transient absolute temperature; ρ is the mass density; C_v is the heat capacity per unit mass object; t is time; and W is the thermal energy generated per unit volume per unit time.

2. Numerical Simulation Method

Except for a very few single and ideal problems, most packaging problems cannot get their displacement, stress, and deformation distribution from the analytical method. A numerical simulation is needed to solve these problems. Among the available numerical tools, the finite element method (FEM) is one of the widely used tools in mechanical analysis.

Presently, some of the design software packages that are widely used assist in the design of microsystems packaging, such as ABAQUS, ANSYS, possess powerful FEM-based calculation modules for mechanical analysis.

Figure 2.13 shows the structural design diagram of a high-temperature package in the power-MEMS design. Metal tubes are sintered together with a silicon wafer by a glass powder binder at a temperature of about 1000°C. This packaging structure can work in a wide temperature range, from room temperature to 600°C. Moreover, the structure is very precise, and the interface precision is within micrometers. Therefore, it is important to study the temperature-dependent displacement and stress by using ANSYS. The main steps include the following:

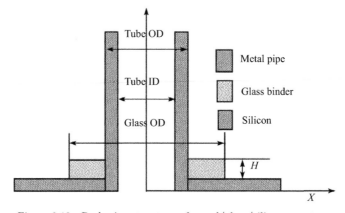

Figure 2.13   Packaging structure of metal/glass/silicon structure

(1) Establish a model and assume the initial and the boundary conditions. The dimensions, thermal-expansion coefficient, Young's modulus, and Poisson's ratio of the studied

packaging structures should be determined in advance. Owing to the axisymmetric feature, the three-dimensional problem can be solved by a two-dimensional axisymmetric model.

(2) A typical mesh is shown in Figure 2.14. For areas with complicated structures, areas with high stress gradients or areas needing specific attention, a denser mesh is required to obtain more precise information.

(3) Calculate the displacement variation in the whole temperature range and work out the corresponding stress distribution. Typical displacement variations and stress distributions are shown in Figure 2.15a, b.

(4) Find out whether the result meets the requirement, and adjust parameters to recalculate if necessary.

(5) Conduct structural design, fabrication, and experimental tests according to the numerical simulation result, compare them with the simulative data and modify the model if necessary.

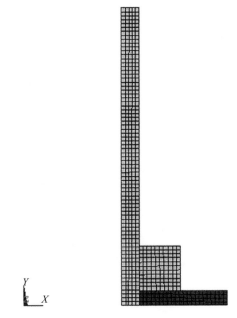

Figure 2.14   Mesh generation

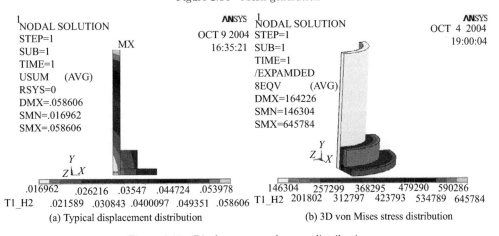

(a) Typical displacement distribution

(b) 3D von Mises stress distribution

Figure 2.15   Displacement and stress distribution

2.4   Microfluidic Design[19−30]

2.4.1   Introduction to Microfluidic Design

Microfluidic design focuses on fluidic problems in the microsystem packaging, such as the design of the microchannel network for the analytical operations in the biochip and the design of the fuel chamber and the nozzle in the microthruster. Most of these microfluidic devices are designed based on the principles of fluid mechanics, and the interface problems are caused by the microscale flow. Although conventional fluid flow has been well studied, and there are a lot of commercial (computational fluid dynamics CFD) software packages for the flow simulation, possible variations of fluidic phenomena induced by size shrinking should be taken into serious consideration in microfluidic design. Typical questions are "What will happen in the microfluidics?" and "Are the equations that have been used to describe conventional fluidic phenomena in the scale of cm–m still applicable in the microfluidics of μm scale?".

2.4.2   Flow in Microchannel

1. Reynolds Number

Reynolds number (Re) is a key parameter to describe the flow status,

$$\mathrm{Re} = vL/v, \tag{2.19}$$

where v is the flow velocity, L is the characteristic length of the flow, and v is the kinetic viscosity of the fluid. Re indicates the ratio of the inertial force to the viscous force, i.e., with the characteristic length decreased, the viscous force affects the flow more and more compared to the inertial force. The flow in a microchannel usually has a very small Re as a result of its micrometer-sized characteristic length; therefore it is a laminar flow.

2. Surface Tension

For the flow inside a microchannel, surface tensions acting on free surfaces will have an important effect on flow performance. For the free surface flow in a straight tube, the pressure drop caused by the surface tension is

$$\Delta P_s = 2\sigma \cos\theta/r, \tag{2.20}$$

where σ is the surface tension coefficient ($\sigma_{\mathrm{water}} = 7.3 \times 10^{-2}$ N/m), r is the radius of the tube, and θ is the contact angle between the liquid and the tube surface. For a water free surface in a 10 μm microtube, the surface tension induced pressure drop, ΔP_s, is about 29 kPa.

3. Hagen-Poiseuille Flow

Generally, an incompressible flow in tube can be described by the well-known Navier-Stokes equations. As a typical example, the fully developed laminar flow inside a straight tube is usually called Hagen-Poiseuille flow. We can derive that the flow rate and the pressure drop in such flow have the following relationship

$$Q = \frac{\pi r^4}{8\mu l}\Delta p, \tag{2.21}$$

and the pressure drop along the tube can be calculated as

$$\left. \begin{aligned} \Delta p &= \lambda \cdot \frac{l}{d} \cdot \frac{\rho v^2}{2} \\ \lambda &= 64/\mathrm{Re} \end{aligned} \right\}, \tag{2.22}$$

where λ is the flow friction coefficient along the tube, μ is the dynamic viscosity, ρ is the fluid density, d and l are the diameter and the length of the tube, and p is the pressure.

4. Increment of the Surface Area to Volume Ratio

When the flow's characteristic length varies from conventional sizes to micrometers, the surface area to volume ratio will increase thousands of times, and this will affect the surface-related heat and mass transfer to some extent. Taking the heat exchanger as an example, the heat transfer area in per-unit volume can be increased considerably by using microfabrication techniques to realize the mini functional structure. In a reported mini heat exchanger, a high heat transfer area of 3600 mm^2 can be realized in a 14 mm^3 volume, and its surface area to volume ratio is about 2.57×10^5m^{-1}. The cooling efficiency is thereby greatly improved.[19]

5. Surface (Interfacial) Force Effects

First, the interfacial force between liquid and solid, the wettability, and the hydrophilic and hydrophobic effects which can be neglected in a designing conventional flow case, should be observed in microfluidic design because of the increased surface area to volume ratio. Furthermore, the downsized flow space will enhance these surface force effects. For example, in some cases, an electrophoresis in a micrometer-sized tube will have a thick, sometimes up to several hundreds nanometers, double electrical layer (DEL). Although this nanometer-sized DEL can be neglected in conventional devices without any question, it will play an important role in microscale electrophoresis. Second, in microfluidics, as mentioned, a large surface area can be easily realized in a small space, so that the inner flow performance can be controlled by tuning the channel's surface properties. For example, the shear flow and the 3D flow inside a duct can be well controlled by a directed electrical field, and this technique has been used to enhance the mixing performance.[23] Moreover, in microfluidic devices, such as the micro blood vessel cleaner, because of the low Re, the surface force effects should be considered carefully since the conventional relations in the high Re flow cannot be applied directly[23−25].

At the same time, the gradient effect should also be considered seriously in the microfluidic design. This is because the size shrinking enlarges some gradients in the flow field considerably, and the flow characters related to these gradients will be enhanced. Most of these microfluidic phenomena need further experimental studies.

2.4.3   Application Example of Microfluidic Package[31]

In most microfluidic applications, connectors are required to connect the chip with the conventional fluidic instruments, such as pumps. A simple vertical connector packaging is schematically illustrated in Figure 2.16a. A drilled hole is used to mount the inlet connector

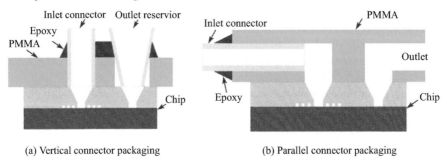

(a) Vertical connector packaging   (b) Parallel connector packaging

Figure 2.16   A typical microfluidic package

with epoxy glue, sealing the gap between the inlet connector and the PMMA (Polymethyl-methacrylate) substrate. Although this packaging method is very simple, it cannot be used to package the chipm which needs microscopy observations due to its large vertical space requirement. A parallel connector packaging, as shown in Figure 2.16b, can solve this problem. The mounting hole is drilled from the side wall of a very thick PMMA substrate. By using this packaging method, the chip can be easily fixed on the carrier and observed by microscopy.

2.5 Multi disciplinary Design

2.5.1 Overview

The challenge of multidisciplinary design always exists in the microsystem. Even in a traditional DIP or more contemporary flip-chip package, multiple fields, such as the mechanical, thermal, and electrical ones, are involved in the manufacturing process and in the device operation afterward. In addition, these fields usually closely interact with each other. With micromachined devices developing toward integrated systems, either in a monolithic or packaged way, the necessity of multi disciplinary designing becomes more and more important, since it may reveal the actual physical, chemical, and biological phenomena and their interaction in device manufacturing and operation. Figure 2.17 conveys the various physical domains (fields) involved and the in-between coupling.

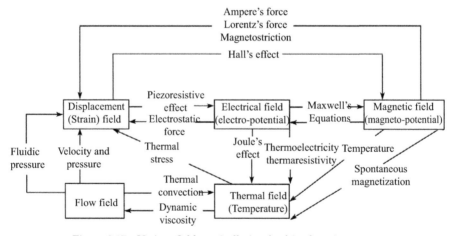

Figure 2.17   Various fields typically involved in the microsystem

The arrows in the figure designate the unidirectional interaction, with the starting point located at the source field, and the final point at the targeted field. For example, the directional lines pointing from "electrical field" to the "displacement field" squares indicates the effect exerted on the displacement distribution by the electro (static) field. The terms in the middle of directional lines denote the physical effects through which the interaction occurs. For example, the "electron static force" between the "electrical field" and "displacement field" squares means that the electrical field inflicts displacement through electrostatic force.

The following conclusions can be reached by analyzing the Figure 2.17. a) bidirectional impacts usually exist between most intercoupled field pairs; b) the thermal field has the widest effective range, since the attributes associated with the entities involved in each field are the functions of temperature; c) almost each of the fields other than displacement may affect this field by the force action, such as the magnetic, electrostatic forces, fluidic pressure, and thermal stresses; d) displacement field and fluidic pressure may exert only weak impacts, and in general the two may have little impact on electromagnetic field; e) the fields with similar natures may easily interact with each other. In all, the above fields can be categorized into three major classes: a) structural and fluidic fields, b) electromagnetic fields, and c) thermal fields.

2.5.2   Multidisciplinary Designing Methodologies

It can be a much more complex task to design a microsystem, including system-in-package (SIP), in a multidisciplinary context, than to do so in a unidisciplined way as mentioned before. In a multidisciplinary design, considering the interactions between variables in fields of different physical natures becomes necessary, as well as analyzing the dynamics of individual fields. For example, for an integrated microfluidic system in which microfluid and electrical signals coexist, the package designing can be quite a challenge in that the interconnect between fluidic variables and electric signals is a requisite for electro microfluidic devices,[32] and the design is in fact a multidisciplinary one.

The multidisciplinary nature of microsystems may complicate the design circles. And the nonlinear correlations between the disciplines involved can further aggravate the analysis challenges that a microsystem package developer has to face. However, a multidiscipline approach can provide a deeper understanding of the underlying physics of a package scheme for microsystems.

Analyses of coupled physical fields has existed in ordinary engineering, and most of them involve coupling between dual fields. General-purpose engineering designing software, such as ANSYS[33] and ABAQUS,[34] provides corresponding modules for fluidic-structural, thermal-structural, acoustic-structural, and force-electrostatic issues. Coventor, Inc.[35] developed the coupled analysis modules based on commercial finite element analysis (FEA) kernels and provides top-down design solutions starting from system-level simulation, which manifested itself as a powerful tool for the analysis of multiphysics in current microsystem packaging.

However, the internal interconnect for energy and substances of various natures is three-dimensional and reticular, and the boundary conditions differ from ordinary engineering issues. Besides, owing to the limited internal space, the multiphysics fields may couple closely with each other, and thus the analysis methodology must be largely different from that of common engineering issues. Therefore, it is necessary to implement iterative analysis and simulations based on available software and to find our a more sound design, analysis methodology, and optimization flow. In this practice, the programming and parametric analysis capabilities may be used to form the integral analysis and simulation solution for multiphysics.

Moreover, a unified approach is more preferred over ad hoc ones. Thus, the eventual buildup of a software specialized for the development of microsystems based on advanced system-level packaging shall lay firm groundwork for the microsystem flourishing.

Multidisciplinary design must also be carried out in a hierarchical way. Starting from system-level modeling is a good option. Though it may be challenging to solve the analytical equations, the mathematical model may reveal the interaction between various fields in a concise and in-depth way.

A top-down approach from the system-level modeling and analysis perspective requires solid modeling to demonstrate physical procedures. The approaches in common use include a) (FEA), which is mostly suitable for the linear and nonlinear analysis on stress and strain field, fluidic fields, and electromagnetic fields, which may be suitable for the simulation both in time and modal domain; b) boundary element method (BEM), which may realize the electrostatic field analysis with adaptive meshing; c) finite volume method (FVM), which may be used not only for the simulation on static (or transient) flow, thermal conductivity, multiphase flow, biological actions, and chemical actions, but also for the static or dynamic analysis on electro-magnetic fields. These approaches may be synthetically used, and corresponding modules can be called upon for specific domain. The modules may be connected through boundary or other coupling conditions (such as the elements of the meshed models), and iteration between individual fields is needed for the solution of the overall performance.

2.5.3   The System-level Simulation by Building Macromodels

Numerical algorithms, like FEM, BEM, and FVM, can be used for individual fields, and appropriate iterations are needed between the coupled fields to obtain system performance; the methods have high accuracy, but with the cost of large and time-consuming calculations. In practice, designers may not painstakingly search for the local features of the system, but instead focus on overall performance, and input-output properties; therefore, macro models may be used for the simplified analysis of coupled fields and the reduction of the degree of freedom. There is no universal method for the buildup of the macro model. The three methods in common use include:

(1) NODAS (node analysis). The system under consideration is regarded as consisting of multiple basic blocks of identical energy domain or various energy domains, with each block acting as a node, just like the fundamental elements in circuitry, such as resistor and capacitors. By using the VHSI Hardware Description Language (VHDL) languages to unite the nodes with practical circuits to form a network, the differential equations may be constructed and simulated by calling SABER or SPICE. The NODAS method has been developed for micro actuators and sensors and described in a lumped element way, so it is not sufficient for complex microsystems.

(2) Black-box model. Realizing the analysis in various energy domains and selecting a few parameters to depict the system energy will greatly reduce the degrees of freedom, without stressing the localized structure and properties, and thus the coupled field will be converted into a black-box model. The steps are a) reduce the degrees of freedom, b) construct the macromodel of the system, c) convert the dynamic equation into A-HDI, which is inserted into analog circuit simulators as a black box. As long as the complete set of deformation and the energy expressions of the system are rationally constructed, calculations with extremely high precision can be obtained. However, the localized features cannot be taken into consideration, and the effect of the structural dimensions of individual components on the system performance cannot be explicitly revealed, which may not facilitate the design and will come with a large calculation amount.

(3) The VHDL-AMS method is to set up a set of normal differential and algebraic equations depicting the component dynamic properties, on the basis of the law of energy conservation. It models large quantities of fundamental components to form corresponding library elements, so that the existing system-level simulator Saber can be used for the simulation on the micromachined elements in tandem with circuitry. If the boundary conditions of the device vary, the VHDL-AMS source code of corresponding devices may be immediately modified and simulated. However, VHDL-AMS is only capable of modeling with normal differential and algebraic equations and is useless for the analysis that involves partial differential equations representing the dynamics of some devices.

The readers may refer to the literature[32,36] for further reading on typical case studies of multi disciplinary design, and to the literature[37−39] for an in-depth analysis of multi disciplinary modeling and simulation methodologies for engineering problems.

Though various methodologies have been presented above, the unified and top-down designing, modeling, and simulation solutions for the closely intertwined fields involved in a highly integrated microsystem or SIP are yet to come. However, with the continuous increase in the calculation power of PCS and workstations and the accumulation of relative knowledge, powerful, versatile, and unified tools will be available in the near future for developers assigned to microsystems or SIP designs.

Questions

(1) What does the design principle for microsystem packaging mainly contain?

(2) What are differences between the design techniques for microsystem and microelectronics?

(3) Please describe the design tools widely used for microsystems and give a regular design procedure.

(4) Give the major steps for the multidisciplinary design of a specific microsystem, such as an accelerometer/inertial measuring unit.

References

[1] R. R. Tummala. Fundamentals of Microsystems Packaging. New York: McGraw-Hill, 2001.

[2] J.H. Lau, C.P. Wong, J.L. Prince, et al. Electronic Packaging: Design, Materials, Process and Reliability. New York: McGraw Hill, 1998.

[3] H.B. Bakoglu. Circuits, Interconnections and Packaging of VLSI. Boston: Addison-Wesley, 1985.

[4] W.D. Brown. Advanced Electronic Packaging—With Emphasis on Multichip Modules. IEEE Press, 1998.

[5] H.W. Johnson and M. Graham. High Speed Digital Design: A Handbook of Black Magic. Upper Saddle River, NJ: Prentice Hall, 1993.

[6] D.M. Pozar. Microwave Engineering. Boston: Addison-Wesley, 1990.

[7] S. Rosenstark. Transmission Lines in Computer Engineering. New York: McGraw-Hill, 1994.

[8] S. Senthinathan. and J.L Prince. Simultaneous switching ground noise calculation for packaged CMOS devices. IEEE Journal on Solid-State Circuits, 26.11 (1991): 1724–1728.

[9] R.R. Tummala and E.J. Rymaszewski. Microelectronics Packaging Handbook. London: Chapman and Hall, 1997.

[10] A. Kraus. The Use of Steady State Electrical Network Analysis in Solving Heat Flow Problems. 2nd National ASME-AIChE Heat Transfer Conference. 1958.

[11] P.I. Frank and P.D. David. Fundamentals of Heat and Mass Transfer. Fourth Edition. New York: John Wiley and Sons, 1996.

[12] W.H. McAdams. Heat Transmission. New York: McGraw Hill, 1954.

[13] J. Howell. A Catalog of Radioactive Transfer Factors. New York: McGraw Hill, 1982.

[14] Luciana W. da Silva and M Kaviany. Miniaturized thermoelectric cooler. Proceedings of IMECE'02, 2002, 1–15.

[15] A. Bergles, A. Bar-Cohen. Advances in Thermal Modeling of Electronic Components and Systems, New York: ASME Press, 1990.

[16] J. Lau and G. Barrett. Stress and De ection Analysis of Partially Routed Panel for Depanelization. IEEE Transactions on CHMT, 10.3 (1987): 411–419.

[17] A.P. Boresi, O.M. Sidebottom. Advanced Mechanics of Materials. New York: John Wiley and Sons, 1984.

[18] J.H. Lau. Creep of Solder Interconnections under Combined Loads. IEEE Transactions on CHMT, 8 (1993): 794–798.

[19] K. V. Sharp, R. J. Adrian and J. G. Santiago, et al. Liquid ows in microchannels. Personal Communication, 2001.

[20] Minglun Xue, Zhanhua Li, Some limitations of the digital micro propulsion miniature. Micro-Nanometer Science and Technology, 5 (2000): 125–127.

[21] G.H. Mohamed The uid mechanics of microdevices—the Freeman scholar lecture. Journal of Fluids Engineering, 121 (1999): 5–33.

[22] A.D. Stroock, M. Weck. D. T. Chiu et al. Patterning electro-osmotic ow with patterned surface charge. Physical Review Letters, 84.15 (2000): 3314–3317.

[23] Yong Li, Min Guo, Zhaoying Zhou, et al., Micro electro discharge machine with an inchworm type of micro feed mechanism. Chinese Journal of Scienti c Instruments, 17 (1996): 56–60.

[24] P. Gravesen, J. Branebjerg, and O.S. Jensen. Micro uidics–a Review. Journal of Micromechanics and Microengineering, 3 (1993): 168–182.

[25] Xiaoning Jiang, Yong Li, Zhaoying Zhou, et al. Experimental Study on Flow Behaviour of Fluid in Micro-Pipe. Proceedings of International Symposium on Manufacturing Science and Technology for the 21st Century (MST'94), (1994): 118–122.

[26] J. Pfahler, J. Harley, H. Bau, et al. Liquid Transport in Micron and Submicron Channels. Sensors and Actuators, A21–A23 (1990): 431–434.

[27] Makihara Mitsulhiro, Sasakura Kunihiko, Nagayama Akira. The ow of liquids in micro-capillary tubes: consideration to application of the Navier-stokes equations. Journal of the Japan Society of Precision Engineering 59 (3), (1993): (399–404)

[28] Xiaoning Jiang, Zhaoying Zhou, Yong Li, et al. Study on Micro uid Flow Behaviour. Optics and Precision Engineering, 1995, 3(3): 51–55.

[29] Zhanhua Li and Haihang Cui. Characteristics of Micro Scale Flow. Journal of Mechanical Strength, 4 (2001): 476–480.

[30] Xiaoning Jiang, Zhaoying Zhou, Yong Li, et al. Study on Micro uid Flow Behaviour. Chinese Journal of Scienti c Instruments. 1995, 16(1), 346–350.

[31] D.S. Xu, A Primary Study on the Chip Preprocessing Blood Sample for Bio-MEMS Application, Master, thesis, Beijing University, 2007.

[32] Paul Galambos and Gil Benavides of Sandia National Labs. Electrical and Fluidic Packaging of Surface Micromachined Electro-Micro uidic Devices. Proceedings of SPIE, 2000 (4177): 200–207.

[33] www.ansys.com

[34] http://www.simulia.com/

[35] http://www.coventor.com/

[36] Luciana W. da Silva and M. Kaviany. Miniaturized thermoelectric cooler. Proceedings of IMECE'02, 2002, 1–15.

[37] C. A. Felippa, K. C. Park and C. Farhat. Partitioned Analysis of Coupled Mechanical Systems [J]. Computer Methods in Applied Mechanics and Engineering, 2001, 90 (24–25): 3247–3270.

[38] C. Boivin, and C. Ollivier-Gooch. A Toolkit for Numerical Simulation of PDEs II. Solving Generic Multiphysics Problems [J]. Computer Methods in Applied Mechanics and Engineering, 2004, 193 (36–38): 3891–3918.

[39] Michael A. Weaver. Nonlinear multi-discipline analysis of conjugate heat transfer and uidic-structure interaction, Ph.D. Dissertation, Georgia Institute of Technology, September, 1997.

CHAPTER 3

Substrate Technology

3.1 Introduction

A substrate is a board onto which a number of individual electrical components, devices, and modules are integrated into functional electronic systems. It is a critical part of microsystems packaging. The ongoing development of IC chip technology and assembly technology has resulted in increasingly high requirements for the performance of substrates. Challenges faced by substrate technology mainly come from the following three areas: (1) Development of microelectronic devices. Microelectronic devices with larger area, quad flat package, surface mount, array pin, I/O, and finer lead pitch are becoming the trend; (2) Development of passive components. Leadless, miniaturized, SMC technology is required to design, to fabricate together with the substrate, and furthermore to be buried into the substrate. (3) Applications in microsystems. Advanced applications prefer substrate with higher wiring density, finer interconnection between layers, and three-dimensional structure.

Substrates used in microsystems packaging mainly falls into the following three categories:

(1) Organic substrates, including paper substrate, woven glass substrate, composite material substrate, epoxy-resin substrate, polyester resin substrate, or heat-resistant plastic substrate, flexible substrate, multilayer wiring substrate, and so on.

(2) Inorganic substrates, including metal substrate, ceramic substrate, glass substrate, silicon substrate, diamond substrate, and so on.

(3) Composite substrates.

Substrate selection and design requires consideration of a number of factors, mainly including material parameters, electrical parameters, thermal parameters, and configuration parameters.

(1) Material parameters include electrical permittivity, coefficient of thermal expansion, thermal conductivity, and so on.

(2) Electrical parameters.

(I) To reduce T_{pd} (time of propagation delay) of substrates, lower electrical permittivity is required.

(II) Matches of various characteristic impedances of different parts within a system should be considered.

(III) To reduce parasitic L, C, and R, it is necessary to minimize leads spacing, use substrate materials with low magnetic conductivity and low electrical permittivity.

(IV) To reduce cross-talk noises, avoid parallel wiring with overdense lines and use of substrate materials of low electrical permittivity.

(V) Prevent signal reflection noise with proper circuit patterns.

(3) Configuration: finer wiring patterns, smaller through-holes for interconnection between layers, and optimization of different electrical parameters are required.

(4) Thermal parameters: key parameters, including thermal-resistance and thermal expansion coefficient of the substrate, should be considered, for example, good thermal match with chip materials such as Si.

3.2 Organic Substrate

3.2.1 Introduction

Organic substrates feature low electrical permittivity, a simple fabrication process, and low cost. Organic materials for the substrates mainly include FR-4 epoxy-glass, BT epoxy, and polyimide and cyanate. Printed circuit board (PCB) fabrication techniques have been widely deployed. Almost every electronic product contains a PCB, on which various components of different sizes are mounted. Major functions of a PCB are to mechanically fix and electrically interconnect all the components and devices.

Increasingly sophisticated electronic systems call for more and more components, which leads to denser and denser wiring on the board. A bare board with no components on either side is usually called a printed wiring board (PWB), which is also widely used in the international packaging industry instead of PCB.

Generally, a core plate of an organic substrate is made up of a thermal-insulating, strong material. A rigid layer of metal foil (mainly copper foil) tracks on the outer surfaces, electrically interconnecting or electromagnetic shielding all components on the PCB.

PCBs are ordinarily green or brown in color as a result of a layer of the solder mask coating on the surface. The solder mask serves as an insulator, prevents copper oxidization, and also prevents components from being incorrectly soldered and connected to metals that will lead to short circuit.

3.2.2 Fabrication Processes of PWB

According to the approach of making metal foils, usually, the fabrication processes of PWB can be classified into two categories: an additive process and a subtractive process. For the additive process, a wiring pattern is deposited by coating conductive materials onto the insulating surface of the board. For the subtractive process, a wiring pattern is formed by removing some needless parts from copper foil, which is coated on the substrate before the process, with chemical etchant. There are two versions of electroplating in the subtractive process, namely panel electroplating and pattern plating.

The panel electroplating approach has the following advantages: first, the electroplating ensures even distribution of dense wiring to prevent sparse wires overloaded by current flow; second, there is no need to adjust the plating current with the area of circuit patterns as in pattern plating, which is better for mass production and thus has found more practical applications;[1] third, there is no special requirement that the photoresist be used, because of the etching-resistant layer formed by the lithography process. The disadvantage of the approach is that it is hard to achieve finer patterns by panel electroplating because of the thick copper foil to be etched.

As shown in Figure 3.1, the basic process is as follows:

(1) Via-holes are drilled in the substrate laminated with a double-sided copper layer.

(2) An electroless copper layer is plated onto the surface of via-holes for electrical connection.

(3) Another copper layer is electroplated over the copper seed layer.

(4) Plate is coated with photoresist.

(5) Plate is exposed and developed.

(6) After copper layer is etched to obtain the required copper pattern, the photoresist is removed.

The pattern plating approach also starts with drilling holes on the board, and then a copper layer is chemically deposited inside the holes for electrical connection. Unlike the panel electroplating approach, the process of pattern plating creates wire interconnections through the following steps, first, coating photoresist on those areas that do not require metallic pattern, second, electroplating copper and other metallic protective layers, such as tin or Sn-Pd solder layer, on the remaining area for the traces and pads of the circuit pattern, and finally, removing photoresist. This process also has pros and cons. The photoresist can prevent the deposit of metal under the photoresist pattern. Finer features can be obtained through pattern plating, such as a smooth surface and ordered wiring. However, it requires that the process parameters should be adjusted to the changing surface or design rule of the conductive area in a timely manner. Another problem lies in the uneven thickness of the plated surface on the same board.

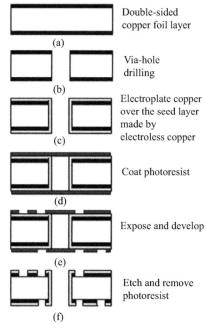

Figure 3.1   Process of panel electroplating approach

According to the numbers of conductive wiring layer, commonly used organic substrates mainly fall into the following categories: single-sided boards, double-sided boards, and multi-layer boards.

1. Single-sided Boards

For the simplest PCB assembly design, devices and components are mounted on one side with wiring patterning on the other side. It is called a single-sided board, as the wiring pattern is fabricated on one side of the board only. Wiring design based on the single-sided board cannot include cross wiring. Therefore, this kind of board was used mainly in early electronic products. Configuration of a single-sided board is shown in Figure 3.2.

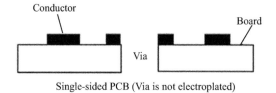

Single-sided PCB (Via is not electroplated)

Figure 3.2   Configuration of a Single-sided Board

2. Double-sided Boards

To ensure more flexible and easier electrical interconnection, double-sided boards were developed, and they have conductive patterns on both sides, as shown in Figure 3.3. Patterns on the two sides need to be connected through a "connection bridge," which is called "via." A via is a vertical hole on a PCB filled or plated with metal foils, through which the wiring patterns on both sides can be connected. Since the area of double-sided boards is twice that of single-sided boards and wire crossings are allowed, they are widely used in applications with high-density components, such as computers, communications, office automatic equipment, and so on.

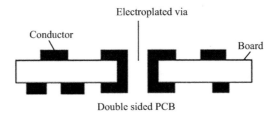

Figure 3.3   Configuration of a double-sided board

3. Multilayer Boards

The development of high-density packaging (HDP) requires increasingly high-density wiring that cannot be accomplished on single-sided or double-sided boards. Today, multi-layer boards are widely used to obtain larger wiring area and more flexible wiring rules. A multi-layer board contains more than two double-sided boards that adhere to each other by means of an insulating layer in between. The layer number of multilayer boards is normally equal to the number of independent wiring foils plus the two outer boards. Generally, the number of layers is even.

To meet the requirements of high-density interconnect (HDI), 3D interconnection including through holes, blind vias, or buried vias should be designed and fabricated within every board. The blind via connects an outer layer with an inner layer next to it. A buried via realizes the interconnection between any two layers.[1]

However, it is a relatively complex process to obtain HDI inside the multilayer board through vias, blind vias, and buried vias. The process will involve repeating a number of steps including laminating, drilling, metallization of the vias, and pattern plating of each layer. Taking an eight-layer PCB as an example, a typical fabrication process of a multilayer PWB is illustrated here:[2]

As shown in Figure 3.4, the board is as thick as 0.8 mm. Blind vias exist in layers L1−L4 and L5−L8, and shallow blind vias in L1−L2 and L7−L8. In addition, there are wiring patterns on L4 and L5. Therefore, two four-layer boards are produced first and then pressed together. Detailed processes are as follows:

The process starts with inner layer etching and exposing; forming inner layers of L2–L3 and L6–L7; inner layer AOI; blackening; laminating four-layer boards of L1–L4 and L5–L8; drilling blind vias on L1–L4 and L5–L8; via metallization; patterning the film affixed to the outer surface of L4 and L5 to form wiring patterns; developing; pattern plating; etching the outer surface; outer layer automated optical inspection (AOI) of L4 and L5; inner layer blackening; laminating L1–L8; punching holes; patterning the film affixed to the outer surface layer at the locations of blind vias; developing and etching the film; AOI; punching blind vias on L1–L2 and L7–L8; via metallization; exposing films affixed to the outer surface of L1 and L8 to form wiring patterns; developing; pattern plating; etching; L1 and L8 AOI; screen-printing solder mask; label printing; plating gold; electrical test; packaging.

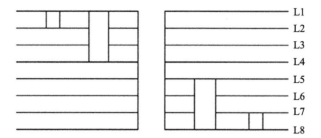

Figure 3.4 Typical configuration of an eight-layer PWB

The AOI system is applied to inspect the process defects and to minimize failures of PWBS by taking immediate corrective actions or improving the PWB fabrication processes.[3]

3.2.3   Application of PWB Substrates in MCM

A multichip module (MCM) is a specialized package where chip dies or CSP components are mounted on one substrate. The substrate can be a PCB, a thick/thin film ceramic substrate. or a silicon wafer with interconnected patterns.[4] A MCM based on a multilayer laminated organic PWB of HDI is referred as on MCM-L, which is mainly used in electronics applications with frequencies below 30 MHz. An MCM-L usually has 6–8 layers or 10–12 layers, two of which are signal layers, and the rest are power layers, ground layers, soldering pads, or distributing layers. Lead pitch is usually within 70–120 μm. A via often has a diameter between 300 and 450 μm. A soldering pad typically has a diameter of 200–300 μm. Substrate materials mainly include FR-4 epoxy-glass, BT-epoxy, ployimide, and cyanate. Substrate fabrication is similar to the normal PCB fabrication process, except that MCM-L is required for etched conductive lines 75 μm wide and vias 150 μm in diameter through laser-drilling or tripping technology. A single polyester substrate together with a 100 μm laminate should be controlled within 1500 μm in thickness. Wire bonding is done by thermosonic bonding, and leads pitch can be as narrow as 40 μm.

Based on mature fabrication technologies, the advantages of MCM-L are they cost liffle, they are capable of mass production, and they provide copper conductive layers of various thicknesses. It can also be used for assembling components on both sides of the substrate. However, the problems to be further resolved mainly include chip inspection for quality control, thermal conductivity of thin laminates, and coefficient of thermal expansion (CTE) mismatch between chips and PCB substrates.[5]

3.3   Ceramic Substrates

Compared with organic substrates, ceramic substrates are superior for their resistance to heat, high thermal conductivity, proper CTE and being ease of use for finer wirings. Therefore, ceramic substrates have been widely applied in packaging for large-scale integrated circuits (LSIs) and hybrid integration circuits (HICs).

3.3.1   Classification and Fundamental Properties

Currently, most widely used materials in ceramic substrates mainly include Al_2O_3, BeO, SiN, mullite, AlN, and glass ceramics, which fall into the following two categories:

(1) Low permittivity substrates, such as, Al_2O_3, glass ceramic substrates and so on. They are easy to fabricate multilayers and mainly used in packaging high-speed devices.

(2) High thermal conductivity substrates, typical materials are AlN and BeO substrates. They are mainly used in assembling power components.

To ensure high performance, ceramic substrates should have high thermal conductivity and a CTE similar to the chip material such as Si. With the increase of integrated density of power chips, power consumption is going up noticeably and chip size is also expanding. Therefore, electronic power devices usually prefer ceramic substrates of better thermal conductivity and proper size and cooling structure to meet the requirements for packaging large-scale and high-power chips. High thermal conductivity substrate will remove more heat from working devices, which ensures high reliability and long lifetime of the devices. The CTE of the substrate should also be close to that of the silicon wafer, which is around $3 \times 10^{-6}/°C$, so as to prevent failures resulting from excessive thermal stress between chips and substrates during packaging and assembling processes. This is especially essential for packaging LSI chips.

The basic requirements of properties for ceramic substrates are as follows.[6]

(1) Electrical properties: low electrical permittivity, low dielectric dissipation, high insulation resistance, high breakdown voltage, and stability in a high-temperature and high-humidity environment.

(2) Thermal properties: high thermal conductivity, proper CTE close to that of the device to be mounted, and good heat resistance at high temperature.

(3) Mechanical properties: high mechanical stiffness, easy to process, good manufacturability for fine pitch and multilayer process, no distortion, no warp, and no flaws cracks and so on.

(4) Other properties include.

    (I) good chemical stabilities and ease of metallization.

    (II) low moisture absorption.

    (III) nontoxic and nonpolluting.

    (IV) low cost.

3.3.2   Fabrication of Ceramic Substrates

There are two typical processes to fabricate ceramic substrates as introduced below: tape casting and powder pressing.

1. Tape Casting

One process of making superior ceramic substrates, known as tape casting, starts with continuously pouring paste tape in uniform thickness on a moving baseband under a scraping blade, after vacuum deformation, which creates a suspension paste made of ceramic powder, plasticizing agent, solvent, and dispersant. After drying, the soft tape, known as a green sheet, will become high-quality ceramic substrate through precutting, organics burning out, and sintering steps. The tape casting is widely used for large-scale fabrication of ceramic substrates in LSI packaging and HICs because of its efficiency in producing multilayer structures. Figure 3.5 shows the process of tape casting in making a green sheet for substrate fabrication.[6]

Preparation of the ceramic slurry is essential during the tape casting process. Granularity of the material is critical to the structure and performance of the ceramic slurry: Smaller particles will benefit from sintering a uniform and compact substrate by improving the osculant reaction probability among particles, the plasticity, and moulding of the slurry. In addition, the conformation of particles guides the thickness of the cast sheet.[7,8] To meet various requirements in different applications, ceramic powder needs to be mixed with 0.5% to 8% additives. For Al_2O_3 ceramic, common additives include oxides such as MgO, SiO_2, and CaO. These additives will help to lower the sintering temperature and make the metallization easier. Furthermore, a certain quantity of organic binder and solvent should be added for making the slurry via the ball milling process. In order to ensure the slurry flows well, surfactants should also be added.

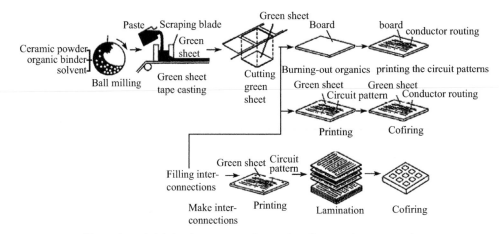

Figure 3.5   A fabrication process of ceramic substrates by tape casting

Ceramic slurry passes through the blade of the tape casting machine onto an organic substrate at a stable speed to form a tape with uniform thickness. After drying, a soft green sheet is formed. After cutting the green sheet and punching vias and cavities, various kinds of substrates can be fabricated according to the three processes as follows.

(1) Lamination, hot pressing, burning-out organics, printing the circuit patterns, and sintering. Circuits can be made by thick-film or thin-film techniques. This process is very flexible, with adjustable parameters, noble metal of low melting point, and no special requirement for sintering atmosphere. This process is mostly used in packaging LSI and HICs.

(2) Lamination, printing the circuit patterns, hot pressing burning-out organics, cofiring. Since the ceramic substrate and circuit patterns are cofired at a high temperature, for example, the sintering temperature for Al_2O_3 is between $1500°C$ and $1600°C$, it requires conductor metals of high melting point, such as Mo and W. To prevent oxidization, sintering should be conducted in a protective and deoxidizing atmosphere such as nitrogen or hydrogen. Generally, the electrical resistance of metal sintered will be higher.

(3) Printing the circuit patterns, lamination, hot pressing, burning-out organics, cofiring. The sintering atmosphere should be the same as stated in (2). This process is currently one of the major methods for fabricating multilayer ceramic substrates.

2. Powder Pressing[9]

Powder pressing, also known as mould pressing or isostatic pressing, involves processing ceramic powder into components or roughcasts to be sintered to certain sizes, shapes, density, and stiffness.

As shown in Figure 3.6, powder pressing mainly involves compacting the metal powder into the desired shape with the axial movement of multiple plungers in the mould cavity. The pressure applied generally is between 350 MPa and 700 MPa. High pressure leads the power to mechanically interlock, followed by cold welding to get the roughcasts, which are then fired into the shape of the final products under high temperature ($1120-1200°C$) for 30 minutes to 120 minutes.

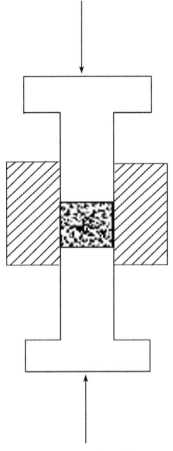

Figure 3.6   Process of mould pressing

There are some limitations to the powder pressing process:

(1) The axial movement of plungers limits the shape of casts expected.

(2) The existence of friction may lead to non uniform density of the casts.

(3) Nonuniform density of roughcasts may cause size changes after sintering.

(4) The mechanical property of the sintered components may be influenced by partial density changes.

3. Pattern Formation on Ceramic Substrates[6]

Circuit patterns are formed by the metallization of the surfaces of a ceramic substrate, which facilitate interconnections between carriers and I/O terminals. Ceramic substrate metallization techniques include:

(1) Thick-film processing, during which circuits and soldering pads are obtained by printing conductive and resistive pastes on a ceramic substrate with screen-printing process and then cofiring them.

(2) Thin-film processing facilitates metallization on a ceramic substrate by deposition of a metal layer in a vacuum, such as vacuum evaporating and sputtering. In principle, any metal can be deposited on any substrate through these vapor deposition techniques. In multi-layer structures, those thin films that cohere to the ceramic substrate directly are generally made of metals with strong reactivity and cohesion, such as Ti, Zr, Cr, Mo, and W. Then ductile metals, such as Cu, Au, and Ag, are plated, since these materials feature high electrical

conductivity, strong resistance to oxidization, and ease of relaxing thermal tension caused by an unmatched CTE. When a thicker film pattern is expected, electroplating will be used to increase the film thickness.

(3) Cofiring, which involves screen-printing thick film on a ceramic substrate with the metal pastes, such as Mo and W, and then burning-out organics and cofiring them with the substrate to form a substrate product. This technique is also applied to the fabrication of multilayer substrates. The metal parts of the sintered ceramic-metal composite will be plated with Ni and other metals for wire bonding and soldering with LSI chips or fabricating I/O leads and microsolder balls.

Currently, those substrates used in packaging LSI and HICs, especially multilayer substrates, are mainly fabricated by the cofiring technique, which features the following:

(I) Ease in obtaining multilayer structures with finer conducting patterns and high-density wiring.

(II) Integrative packaging structure of insulator parts and conductor parts can provide a solution of hermetic package.

(III) Capable of controlling the shrinkage rate after the sintering process by selecting the desired elements, molding pressure and, sintering temperature.

3.4   Introduction of Typical Ceramic Substrates[6]

There are many types of ceramic substrates. The most widely used ones in electronic packaging mainly include Al_2O_3, mullite, Luow-temperature cofired ceramics (LTCC), BeO, AlN, and SiC substrates. The essential properties of these ceramic substrates are listed in Table 3.1.

Table 3.1   The properties of widely-used ceramic substrates

Property	99%Al_2O_3	Mullite ceramics	LTCC	BeO	AlN	SiC
Flexural strength (MPa)	290	140	130$-$300	170$-$230	400$-$500	450
Breakdown voltage (kV/cm)	150	130	200$-$500	100	140$-$170	1$-$2
Dielectric constant ε_r (1/MHz)	9.8	6.5	4.2$-$8.0	6.5	8.8	45
Dielectric loss tan δ (1/MHz)/$\times 10^{-3}$	0.1	0.4	0.5$-$0.3	0.5	0.5$-$1	50
CTE (25°C)/(10^{-6}/°C)	6.8	4.0	4$-$6	8	4.5	3.7
Thermal conductivity (25°C)/(W/(mK))	31.4	4.19	3$-$8	250	100$-$270	270
Sintering temp. (°C)	1500$-$1650	1400$-$1500	<900	2000	1650$-$1800	2000

3.4.1 Al_2O_3 Substrate

Among ceramic substrates, Al_2O_3 substrates have the best general performance and are the most widely used with relatively low prices. They are mainly fabricated by the lamination molding technique.

Al_2O_3 ceramics, with thermal conductivity of around 0.2 W/(cm·K) and electrical permittivity of about 10, are unable to meet the requirements of the development of today's ICs. Therefore, their properties must be meliorated.

The traditional substrate fabrication processes are mainly based on tape casting, a complex process that is hard to control and expensive to implement, which adds to production costs. A 99% Al_2O_3 substrate fabricated by gel-casting, a simple process with low production costs, can be widely used for industrial manufacturing.[10]

Al_2O_3 substrates have high reliability and strong airtightness. They are mainly used in the substrates for HICs, LSI packaging, and multilayer circuits composite structures.

3.4.2   AlN Substrate

AlN substrates have high thermal conductivity, over 10 times that of Al_2O_3 substrates. Their CTE is well matched with that of silicon dies, which is essential to the packaging of high-power semiconductor chips and high-density packaging. AlN substrates used in MCM packaging especially have very good prospects. They enjoy advantages such as high stiffness and low mass density. Their good machinability makes them applicable for both tape casting and sintering in atmospheric pressure. They are considered ideal substrates for the new generation of high-density and high-power electronic packaging.

AlN is man-made material, and AlN powder for commercial use is mainly obtained by two processes, namely, carbon deoxidization with Al_2O_3 and direct nitrification of Al.[11] Compared with the other processes, AlN powder made from direct nitrification of Al has a wider range of scales and a larger average particle size. AlN green sheet can be obtained by ordinary ceramic processes such as dry-pressing, tape casting, and mold casting. In particular, tape casting has the advantages of high productivity, ease of cceation and uninterrupted and automatic production. It can also improve product quality, lower production cost and make mass production possible.[12]

Metallization of the substrate is an important step to get AlN substrates from the research and development (R&D) stage to common use. However, It is difficult to apply metallization to AlN because it is made from strong covalent bond compounds by sintering and it has weak reactivity to other materials and weak wettability.

Directly bounding copper (DBC) on an AlN ceramic surface is a useful technology for metallization, known as AlN-DBC technology. Substrates fabricated by AlN-DBC technology offer the advantages of high thermal conductivity of AlN ceramics and high electrical conductivity of copper foils, which are suitable for applications in high-power electronic module, DC/CD power module, solid-state relays (SSRs), thermoelectric cooling sets and so on.

DBC technology was originally used in the preparation of Al_2O_3 substrates. Cu foil is bonded onto the Al_2O_3 substrate in N_2 atmosphere containing O_2 with a temperature of about 1063°C. AlN is a nonoxide ceramic that is essential to form a transition layer for bonding copper foil with the substrate. It usually starts with surface treatment. That is, in a normal atmosphere, the substrate is heated to 1050−1075°C and then sintered for 30 minutes. The next step is to place the Cu foil on the treated surface and put them into a high-temperature diffusion oven. Finally, after 0.5 hour to 1 hour of natural bounding in N_2 atmosphere at a temperature of 1063°C to 1070°C, a AlN-DBC substrate is formed.

Various parts of a MEMS device are designed and fabricated in a compact space, and their functions closely influence each other. For example, the circuit parts generate a great deal of heat, and the mechanical moving parts are very fragile. It is important to ensure good transfer of signals among parts and provide effective protective measures. This requires higher standards in MEMS packaging technologies. AlN ceramics are among the ideal materials for MEMS packaging because they have high thermal conductivity ($\geqslant$200 W/(m · K)), low electrical permittivity, high stiffness, similar CTE to Si, and other good physical properties, as well as good chemical stability and strong corrosion resistance.[13,14]

3.4.3   Mullite Substrate

Mullite, consisting mainly of AlO_6 and $AlSiO_4$, is one of the most stable crystalloids in the Al_2O_3-SiO binary system. It has good chemical and thermal stability, favorable electrical properties, and low weight. Compared with Al_2O_3, it has lower mechanical stiffness and thermal conductivity. However, it can be used to further improve the signal transmission speed with its low electric permittivity. In addition, mullite's low CTE can reduce the heat stress of LSI carriers. Also, its CTE is close to that of conductive materials like Mo and W, so that the stress between conductors and the ceramic is very low during sintering. In particular, high-purity and superfine mullite ceramics have been ideal material for VLSI packaging or substrates. Therefore, mullite and mullite-based ceramic glass is getting more and more attention as a high-performance material for packaging.[15]

Mullite ceramics are mainly made from SiO_2 added with glass additives and cordierite. With various compositions of these elements, ceramics of different electrical permittivity and dielectric loss can be obtained.[16] With the reduction of mullite content, the permittivity of the mullite ceramics declines. When the content accounts for 70%, with glass additives a ceramic of permittivity close to 6.0 can be obtained; and with cordierite, not less than 6.5. In the case of high mullite content, permittivity of ceramic with the two kinds of additives is not higher than 7.0. However, the dielectric loss increases as permittivity goes up. Normally, cordierite additives are better than ceramics additives in preparing ceramic for a lower dielectric loss. These indicate that different properties can be obtained by changing the proportion of different additives, which will have notable influence over the development of sintering processes and mullite crystal grains.

Different sintering temperatures can lead to different characterizations of mullite ceramics added with glass and cordierite. These mullite ceramics have a strength ranging from 160 MPa to 220 MPa. At the same sintering temperature, the higher the mullite content, the higher the strength. There is an appropriate temperature corresponding to a required strength.

3.4.4 SiC Substrate

SiC is a composite with strong covalent bond, and its rigidity is comparable to those of diamond and cubic boron nitride (c-BN). In addition, SiC has good wear resistance and strong resistance to most chemical etchants. The most unique property of SiC is its high thermal diffusivity compared to other materials, which is even higher than copper (1.1 cm^2/s). Its CTE is closer to that of Si. The weak points of SiC include: first, high relative permittivity, which is, for example, 40 at 1 MHz and 15 at 1 GHz; second, weak breakdown voltage. SiC substrates will become conductive and be easily broken through when the intensity the electric field reaches several hundred volts per cm.

SiC substrates are mostly used in low -voltage circuits, and packaging VLSI which requires high thermal conductivity.

Al/SiC is an improved type of material based on SiC, which is made by adding the melted Al into SiC to form an alloy under high temperature and high pressure. Al/SiC combines the advantages of the two materials: metal's high thermal conductivity and ceramics' low CTE. It can meet the various requirements of multifunctional characteristics in design as an excellent electronic packaging material integrating a number of properties, including high thermal conductivity, low CTE, high stiffness, low mass density, and low production cost. Internationally, Al/SiC is widely regarded among innovated materials as the third generation of electronic packaging materials, which has great prospects in applications of high-performance microwave circuits in industries such as military and aerospace.[17]

Excellent characteristics of Al/SiC include:

(1) High thermal conductivity (170−200 W·(m·K)$^{-1}$) and adjustable CTE (6.5×10^{-6}− 9.5×10^{-6}). On the one hand, the CTE of Al/SiC matches those of semiconductor chips to prevent failures from fatigue. A power chip can even be directly assembled on the Al/SiC substrate. On the other hand, Al/SiC's thermal conductivity is times 10 as high as Kovar alloy, and the heat generated by chips can be easily dissipated.

(2) Adjustable CTE. Various CTE levels can be obtained by changing the proportions of elements so that a product can be developed according to the user's requirements, an

advantage that hardly any other metal material or ceramic material can offer.

(3) Mass density close to that of Al and much less than that of copper and Kovar, accounting for less than one-fifth of that of Cu/W alloy. Therefore, Al/SiC is especially applicable in portable devices, aviation and avionics, or other weight-sensitive products.

(4) High ratio of stiffness density (stiffness divided by mass density), which is the highest among those of all electronic materials, 3 times as high as that of Al, 5 times as high as that of W-Cu or Kovar, and 25 times as high as that of copper. In addition, Al/SiC has a higher resistance to vibration compared with ceramics. Therefore, it is the first choice for applications in stringent circumstances, such as vibrations that occur in the avionics and automobile industries.

(5) Good airtightness. Al/SiC has strong airtightness itself. However, the final airtight specification of a packaging product with Al/SiC and metal or ceramic depends on proper deposition and soldering processes.

3.4.5   BeO Substrate

BeO has the highest thermal conductivity among all ceramic materials. In addition, it has good electric insulation, low electrical permittivity, and high mechanical strength. Compared with other substrate materials, BeO ceramic is applicable in high-frequency circuits for its low dielectric constant. The thermal conductivity of BeO ceramic is 96%, over 10 times high as Al_2O_3. Therefore, BeO is also applicable in power circuits. Although BeO's thermal conductivity decreases as the temperature increases, its conductivity still exceeds that of AlN by 30%, in the temperature range of 25−300°C. Special chemical bonded or mixed paste is usually used for pattern printing on BeO ceramic substrates with little glass phase, because of its high purity of 99.5%. One of the concerns with BeO is its toxicity, which has been reported in many publications. Because of this, three companies, including the U.S.-based Brush Wellman Inc., announced that solid BeO is nontoxic. BeO particles are only harmful when breathed into human body. However, those particles can be well controlled in the fabrication process. Therefore, it is safe to use BeO ceramics.[18]

BeO substrates are mainly fabricated by dry-pressing. After molding, they are prefired at a temperature between 300 and 600°C, and then fired to form the final product at a temperature of about 1500−1600°C. With this method, the shrinkage in the sintering process is small, and thus relatively accurate size can be obtained. However, it is difficult to control the diameters and spaces of drilling holes to be made in the fired substrate. Another process to make BeO substrates called the green sheet process is done by adding a small quantity of MgO and Al_2O_3. Moreover, it is difficult to control the sizes of particles of the fired substrate, and the particle diameter of BeO is larger than that of Al_2O_3. Therefore, the surface of BeO substrate must be ground and polished before thin-film metallization. Another feature of its metallization is that the combined strength of BeO with Cu is greater than that with Mo or W.

With a number of great properties, such as high thermal conductivity, high melting point, high intensity, high electric insulation, low permittivity, low loss, and good mechanical properties for packaging process, BeO ceramics are regarded as an important material and are widely used in the fields of microwave technology, vacuum electronics technology, nuclear technology, microelectronics, and photoelectron technology. In particular, BeO ceramics are favored in manufacturing parts of high thermal conductivity, which are used in high-power semiconductor devices and circuits, high-power microwave vacuum devices and nuclear reaction piles.[19]

3.5 LTCC Substrates

For Al_2O_3, mullite, AlN, and BeO substrates discussed above, if the metal traces are obtained by cofiring at a temperature of 1500–1900°C, only those metals with high melting points can be selected as the conducting material, such as Mo and W. The following issues will occur:[6]

(1) The deoxidizing atmosphere is needed for the cofiring process, which adds to the complexity of the process. The high temperature requires a special sintering furnace. The high electrical resistivity of Mo and W makes it difficult to use fine pitch wiring, as a result of the high resistance resulting in distortion of signals.

(2) Dielectric materials used have relatively high permittivity, which will increase the transmission delay time. Therefore, they are not applicable in high-frequency circuits.

(3) There is a large CTE mismatch between Al_2O_3 (7×10^{-6}) and Si (3.0×10^{-6}). In cases where a bare die ismountc on Al_2O_3 substrate, high thermal stress will occur during thermal cycling.

To solve these problems, low temperature cofired ceramics (LTCC) have been developed. The sintering temperature is controlled around 900°C, so that those conducting materials with low electrical resistance, such as Au, Ag, Ag-Pd, Ag-Pt, and Cu, can be used for finer wirings. In particular, noble metal paste can be sintered in a normal atmosphere. LTCC substrate offers the following advantages:

(1) It is a high-quality substrate to use in the application of high-frequency circuits. The frequency reaches as high as tens of GHz.

(2) It will be beneficial to improve the quality of circuit systems because of the use of metal materials as conductors with its high electrical conductivity.

(3) It can be used to process fine pitch wiring with a line width less than 50 μm.

(4) It is applicable to circuits with large current flow and extremely high temperature. It also has higher thermal conductivity than ordinary PCB substrates.

(5) It has good thermal performance, for example, low CTE, low permittivity, and low thermal coefficient.

(6) It can be used in the fabrication of multilayer substrates with multiple passive components embedded inside, which helps to increase the packaging density.

(7) It is applicable to a wide range of components and devices. Besides R, L, and C, it can also integrate MEMS devices, sensors, EMI controllers, and circuit protection components on the substrate.

(8) It can integrate various active passive components by wire bonding ICs and active devices on a multilayer 3D circuit board.

(9) It demonstrates high reliablity and endurance in stringent environments, such as high temperature, high humidity, and shock.

(10) It is a nonsequential process allowing inspection into green tapes to improve product quality and reduce the production cost.

(11) It can be formed at low sintering temperatures and uses multilayer metallization and materials of high electrical conductivity, such as Ag, Cu, and Au. In this way, signal loss can be effectively reduced and high-density wiring in multiple layers can be created.

Smaller vias, finer pitch wirings, and higher wiring density are development trends in LTCC technology. Recently, LTCC substrate has been developing rapidly and widely used in more and more areas, including high-performance packaging, high-speed MCM packages, and high-density BGA or CSP packaging.

Another challenge in developing LTCC materials is the new types of ceramics with low R & D costs and better properties. In addition, green ceramic tapes with zero shrinkage and with low loss in high-frequency applications are new products that will be developed very soon.

3.5.1 Properties of LTCC Substrates

Just as its name implies, LTCC should be sintered at low temperatures while keeping other desirable properties. Now, what can be called a "low" temperature? A number of parameters for commonly used conductor materials, such as melting point and resistance, have been listed in Table 3.2. Metals with low electrical resistance include Ag, Au, and Cu, and they all have a melting point below 1100°C, which is considered generally to be the standard of low temperature. When fabricating multilayer substrates with LTCC technology, it is suggested that the sintering temperature be held in the range of 850−950°C.

Table 3.2   **Properties of metals for wirings on ceramic substrates**

Metal	Melt point °C	Boiling point °C	Resistivity ($\mu\Omega\cdot$cm)	CTE (10^{-6}/°C)	Sintering gas
Ag	961	1980	1.6	19.1	Air
Au	1063	2600	2.2	14.2	Air
Cu	1084	2595	1.7	17.0	Deoxidized gas
Ni	1452	3000	7.2	12.8	Deoxidized gas
Pd	1550	2200	10.8	11.0	Air
Pt	1770	3800	10.6	9.0	Air
Mo	2617	4600	5.2	5.4	Deoxidized gas
W	3377	5527	5.5	4.5	Deoxidized gas

The use of metals with low electrical resistance, such as Ag, Au, and Cu, can make finer circuit patterns and high density wirings possible.

In order to meet the requirements of high-speed circuits, the transmission delay time should be shortened, and the permittivity of dielectric materials should be lowered.

In addition, great thermal shocks will be undertaken in attaching large-scale IC bare dies onto the substrate in MCM and under operation conditions in service. Therefore, it is necessary to match the CTE of the substrate materials with that of the IC chip to ensure the reliability of the interconnections.

In addition, the materials must be strong enough to resist various mechanical and thermal stresses in the packaging and assembling processes.

To sum up, basic requirements for the LTCC substrate should include:
(1) Low sintering temperature (<950°C).
(2) Low electrical permittivity.
(3) Similar CTE to that of the attached chips.
(4) High mechanical strength.

3.5.2 Classification and Characteristics of LTCC

(1) Lead borosilicate glass-Al_2O_3 series.

A glass ceramic powder, made from lead borosilicate glass compounded with Al_2O_3 to the ratio of 45%:55% in weight, is sintered at a temperature between 850 and 950°C. In particular, material cofired at the temperature of around 930°C has a maximized flexural strength of 350 MPa, which is similar to that of Al_2O_3. A disadvantage of this material is that it contains lead. It will be replaced by lead-free systems in the near future.

(2) Series of borosilicate glass quartz glass cordierite system.

The weight composition of this series is 65% borosilicate glass, 15% quartz glass, and 20% cordierite. The CTE of this kind of product after the sintering process is 4.4. The glass ceramic material has a low permittivity and a controllable CTE matching that of silicon.

(3) Series of borosilicate glass Al_2O_3 forsterite.

A composition with the highest density in this series is 35% Al_2O_3, 25% forsterite, and 40% borosilicate glass. With this composition, the relative density of the structure cofired at the temperature of 900°C can reach 97%. It has a maximized flexural strength of 200 MPa, and its permittivity is 6.5, while its CTE is $6.0 \times 10^{-6}/°C$.

(4) Series of borosilicate glass-Al_2O_3.

The optimal sintering temperature of this series varies with the content of borosilicate glass. The preferred composition is 50% Al_2O_3 and 50% borosilicate glass. The CTE of the structure cofired at 900°C can reach $4.6 \times 10^{-6}/°C$. And it has a maximized flexural strength of 245 MPa.

(5) Series of borosilicate glass-Al_2O_3 treated ZrO_2.

The optimal composition is suggested as 40%−60% borosilicate glass and 40%−60% Al_2O_3 treated ZrO_2, sintered at 900°C. This material has a CTE well matched with that of GaAs crystal within a quite wide range of temperatures. Therefore, it can be used in packaging substrates for high-electron migratory transistor (HEMT).

3.5.3   LTCC Substrate Fabrication Process[20]

Dielectric materials of LTCC mainly consist of glass/ceramic mixed with other organics such as organic binders and organic solvents. By means of the tape casting process, they are fully mixed and fabricated as high-density green sheet with an accurate thickness. After forming holes on the green ceramic sheet by laser drilling or mechanically punching, metal paste is used to fill interconnections between layers. Then the conductor patterns are printed on each layer by the screen-printing process. After stacking, thermal laminating, or hydrostatic laminating in a sealed bag, laminates are cofired to obtain the multilayer substrate. Figure 3.7 shows the process flow of LTCC fabrication.

1. Tape Casting

The green sheet obtained from tape casting should be flat and of high density. The width of the green film should be not less than 110 mm, and it should be reasonably stiff. The key technologies of tape casting include the mechanical equipment, material composition, and the control of various parameters.

2. Via Fabrication and Filling

Vias can be formed by drilling, punching three to five per second can be holes can be formed by drilling and holes can be over 0.25 mm in diameter with a tolerance of ±50 μm. The drill bit can be easily broken while drilling small holes. The smaller the drill bit is, the more expensive it is. So it is not cost effective to drill holes less than 0.25 mm in diameter with this method. Eight to ten holes per second can be formed by punching, and they can be as small as 0.05 mm in diameter with a tolerance of ±10 μm. Through laser drilling, 250−300 holes/s can be formed, and they can be as small as 0.1 mm in diameter with a tolerance of ±25 μm. For the LTCC manufacturing process, the preferred size of via holes is about 0.15−0.25 mm in diameter, to balance the improvement of wiring density with the process of via metallization. If the size of via holes is ⩾0.3 mm or ⩽0.15 mm, it is difficult to form blind vias with existing metallization approaches, which may decrease the production efficiency and the product reliability.

Via filling is a critical part of LTCC substrate fabrication.[21] Vias can be filled through the following techniques: screen-printing or mask-printing. The work bench of a printer designed for low-temperature processes is made of lacunose ceramic plates or metal boards with a locating pin which is around 1.5 mm in diameter, at each of the four corners, which matches the locating holes on the green sheet. Under the work bench, a vacuum pressure of about 665−886.4 Pa is provided. Screen-printing should use a stainless steel screen with over 250 mesh or a nylon screen with a large coverage of holes.

Contact printing is preferred, with a paste of around 30 μm. A brass or stainless steel mask 25−30 μm thick may be used. A piece of filter paper is placed under the green sheet to prevent the paste from leaking to the work bench through vias. After the printing is completed, the green sheet is taken away together with the filter paper for 5−10 minutes' drying at a temperature of 70−100°C before removing the filter paper. The paste for via filling should have good liquidity and proper viscosity, which should be adjusted to the size of the vias. Generally the viscosity is about 1000−2500Pa · s. Otherwise, it is difficult to form blind vias. After printing, it is important to check the vias under the microscope and to repair imperfect vias.

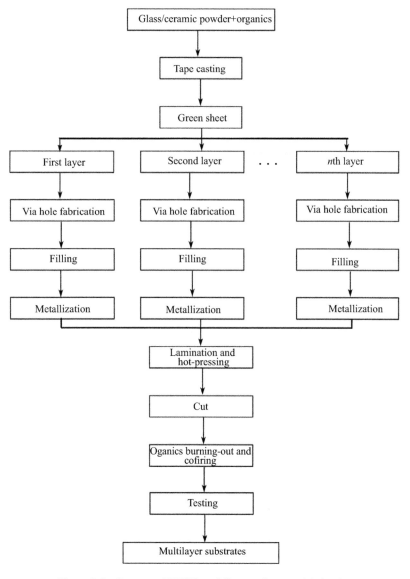

Figure 3.7   Process of LTCC multilayer substrate fabrication

3. Positioning

Positioning refers to the alignment between the screen and the green sheet during printing and between green sheets during laminating. If the positioning tolerance is too wide, open or short circuits may occur. Commonly used tools, such as locating holes or fiducial marks, are used to meet the different requirements in positioning accuracy. For example, when using fiducial marks, it is worth paying attention to the different tolerance requirements in line width, line spacing, via diameter, and via spacing. The positioning accuracy of normal wiring density is about ±50 µm.

Factors that affect the positioning accuracy include drilling tolerance, lithograph mask tolerance, and visual tolerance during manual operations on printers. In order to increase the wiring density, those tolerances must be improved. New printers with automatic vision alignment systems, which can greatly increase the positioning accuracy, are already on the market.[22]

4. Wiring Design and Metallization[23,24]

During wiring design with the CAD tool, the line width and spacing and other parameters must be designed according to the requirements of electrical parameters, positioning accuracy, and sizes of vias. Normally, with the help of LTCC technology, very fine pitch wiring and small spacing can be fabricated, but its process cost is high. A multilayer LTCC substrate with wider wirings is cost effective. During wiring design, both the technology and cost should be taken into consideration. For substrates applied to high-frequency and high-speed circuits, fine wire width and fine pitch should be used. To reduce production cost, it is preferable to choose multilayer structures with wider wire and larger spacing while ensuring the product quality.

Metallization of LTCC substrates falls into two categories: inner surface metallization and outer surface metallization.[25] Screen-printing and computerized direct plotting are the most widely used techniques for inner surface metallization. Besides these two, photolithography and thin-film deposition are also used in outer surface metallization. With the development of finer screens with larger mesh coverage and masks and pastes of higher resolution, lines as thin as 100−150 µm can be screen-printed. Some spacing (called screen spacing, off-substrate distance, or off-contact distance) is kept between screen and substrate, and paste is forced by squeegee at a given speed and pressure along the screen and then is printed on substrate below through the open areas of mesh to transfer images on it. Screen-printing can be classified into two main types, contact mode and non contact mode, by the ways in which the screen returns back to its original position after transferring. The former takes advantage of screen tension to divorce the screen from the substrate and return it to its original position. The latter makes use of mechanical methods to divorce the screen from the substrate, which clings to it during the printing process. In electronics packaging engineering, non-contact screen-printing technology is used most frequently[26] If thinner lines and smaller spacing are required, thin-film deposition or thick-film photolithography can be used, in which lines as thin as 40−50 µm can be obtained. However, these two technologies can only be used for outer surface metallization, and they are quite expensive as well.

Computerized direct plotting is a versatile technique, for it needs no screen. However, it features low throughput, complicated operation, and expensive equipment.

With the increased density and pin numbers of chips attached to the LTCC substrate, smaller width and spacing of interconnection traces are required, as are smoother trace edges, which can hardly be fabricated by traditional thick-film printing technique. This led to the development of thin-film techniques. The process of thin-film metallization is shown in Figure 3.8, and detailed steps are given in the references[23].

Before thin-film metallization, the surface of an LTCC substrate should be ground and polished. Then multilayer films are deposited by the magnetic sputtering process, and soldering pads and conducting traces can be obtained through the lithography process. Finally, substrate products are obtained through heat treatment.

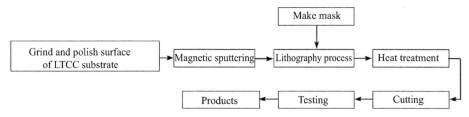

Figure 3.8   Thin-film metallization process of LTCC substrates

5. Lamination and Hot-pressing

After via filling and metallization, the green sheet is shifted into the stack mold with locating pins that will match with the locating holes on the sheet, which ensures positioning accuracy. The best material for the stack mold should have enough hardness to prevent distortion in the process. The laminating and hot-pressing process calls for uniform pressure, which is closely related to the shrinkage of the substrate.

If the pressure is too high during hot-pressing, the substrate will delaminate because of air bubbles as a result of organic burning or out. Delamination will also-occur if the pressure is too weak, together with larger shrinkage and nonuniform shrinking rate of the substrate during sintering process. It is necessary to have in-depth experiments to optimize the pressure of hot pressing. The isostatic lamination machine can provide uniform pressure for idea lamination and equal shrinkage, which is conducive to the improvement of alignment accuracy between sections of soldering pads and between post fire conducting surfaces and through vias.

6. Organics Burning-out and Cofiring

The organics burning-out in a substrate can be carried out in an ordinary muffle furnace. The rate is determined by the thickness of the substrate. A typical condition is that in which the temperature increases at a rate of $0.2-0.5°C/min$ and is maintained for $3-5$ hours when it reaches $450°C$. If the temperature rises too fast, the substrate will delaminate because of air bubbles coming out of the layers. The substrate should be place on a flat ceramic or quartz plate. Besides in a muffle furnace, sintering can also be conducted in a chain furnace with temperature increase at $8°C/min$. When it reaches $850°C$, the temperature should be kept unchanged for 10 min, and then decreased at the speed of $8°C/min$. The key to sintering lies in the accuracy of the sintering curve and the temperature uniformity in a furnace, which is closely related to the flatness and shrinkage of the substrate after sintering. If the temperature inside the furnace is not uniform, the shrinkage of the substrate will hardly be even. During the sintering process, temperature that increases too rapidly may lead to an unsmooth substrate surface and large shrinkage.

An experiment validated that the sintering temperature remained at $850°C$ in both cases. In one case, the temperature ramp rate of $12°C/min$ led to flatness of 120 (µm/50 mm) and shrinkage of 13.8%, and in the other case the temperature ramp rate of $8°C/min$ resulted in flatness of 70 (µm/50 mm) and shrinkage of 13.5%. During sintering, there will be a difference between the temperatures of the conductor and the substrate. If the temperature increases too quickly, stress resulting from different levels of density will cause the substrate to warp. In addition, it may result in melting glass penetrating into the space between ceramic particles, which will increase the shrinkage of the substrate. If the fired substrates are not so flat, an effective way to flatten them is to place the substrate between two ceramic plates forre sintering at the temperature of 850°C.

7. Testing

After the sintering process, the LTCC substrates must be tested to check the performance of the interconnections. Test instruments are mainly the probe testers. For example, the automatic wafer probe tester can measure electrical parameters, such as the capacitor and so on. Two-way probe testers can find failure phenomena such as open circuits, short circuits, and high resistance failures on both sides of an MCM substrate. The minimum testing pad area is not more than 5.08×10^{-2} mm.

The resistor, capacitor, and other passive components can also be fabricated on the green sheet of LTCC substrates. One process is to fabricate these components between the two layers of green sheets. An alternative process, known as "buried components," is to print the passive components on a fired Al_2O_3 ceramic substrate, followed by stacking layers of green sheets on the substrate. Before the stacking process, the electrical resistance of "buried components" printed and sintered on the Al_2O_3 substrate can be accurately adjusted by laser. The pastes developed for this process have high stability even after being sintered many times.

In addition, MEMS structures like microcavity and micropipe can also be produced on the green sheet of the LTCC substrate. It is foreseeable that LTCC technology will play an increasingly important role in the development of future system in package SIM and system on package (SOP) products for its manufacturability. It will become one of the most widely used high-end substrates.

At present, LTCC green sheet technology has already been commercialized. Companies like Dupont, Ferro, and ESL have launched LTCC green sheet products for commercial use. Using commercialized LTCC green sheets to fabricate MCM multilayer substrates can eliminate the extra cost in making stable green tapes to reduce the production cost and shorten the production cycle.

3.6 Advanced Substrates

Next generation portable consumer electronics will require significant improvements of integration and packaging technologies, mainly because of increasing signal frequencies and the demand for higher density of functions at acceptable cost. Recently, more and more researchers are focusing on advanced substrates, such as the substrates with elements embedded, high-density interconnect substrates (HDIS), and flexible substrates, etc., for high density, high reliability, and high frequency propagation required for the next generation semiconductor package.

3.6.1 Substrates with Elements and Channels Embedded

Space requirements of active components can be reduced to a minimum by using chip size packages (CSPs) or flip chips. Further miniaturization, however, requires 3D integration of components. The obvious benefits, to name only a few, are shortened wiring, increased reliability by reduction of solder joints, and better electrical performance by reduction of parasitics, especially for high-frequency applications on the order of several GHz. Embedded passive and active chips or components in substrate has been described to have advantages over classic SMT and LTCC because it offers smaller form factor with a high level of passive integration, thus achieving a lower cost system in package, especially in high-frequency applications. SIP products realize all the system functions on an ultraminiaturized, multifunctional and high-performance microelectronics package by integrating both active and passive components into a single package substrate. Since the early 1990s several embedded passive and active chip technologies have been reported, and a few have been commercialized by universities, research institutes, and companies. For example,the Packaging Research Center (PRC) at the Georgia Institute of Technology demonstrated various substrates with passive and active embedded the for SOP.[27] Imbera Electronics has developed several integrated module board (IMB) technologies for organic, low-cost PCB substrates with discrete elements embedded. Since 2003, Imbera has focused on third-generation technology providing a low-cost, flexible platform for multiple component types, including Si, GaAs, and discrete C, R, in the range of 2 to 350 I/Os[28].

Embedded active technology, in which thinned active chips are directly buried into a core or into high-density interconnect layers, is gaining more interest for ultraminiaturization, increased functionality, and better performance of SIP. The current technology provides organic substrates with high-density build-up layers and microvias, assembled on both sides with surface mount passive and active components.

It is very important to improve the tolerance of embedded elements in order to meet the requirement from high-frequency electronics for stringent tolerance of passive components embedded in a substrate. T. Kim and his colleagues from Samsung Electro-Mechanics Co. Ltd. proposed a new design and process method providing an acceptable level of improvement in each embedded passive component's tolerance by studying mainly the factors coming from the metal layer's line variations. It is reported that resin-coated copper foil type material is used as the embedded capacitor material, while the dielectric material of the embedded capacitor is composite material with ceramic powder and epoxy binder.[29]

The LTCC technology provides a convenient medium for fabricating three-dimensional (3D) structures, such as cavities and channels, and for embedding electrical elements of capacitors, resistors, and 3D interconnections. These structures, capable of being easily embedded or integrated into the substrate for ICs, can become integral components in liquid-cooling systems for packaged ICs, and biomedical analysis (or reaction) systems, and can also be good candidates for the implementation of mechanical sensors and actuators, such as accelerometers and grippers. These functional modules can then largely enhance the performance of those substrates for cutting-edge packages.

For example, chemical sensors, physical sensors and micropumps by Ilmenau University of Technology[30] and thick film accelerometers by Dresden University of Technology were fabricated by using LTCC technology,[31] while the internal microchannels have been used for fluidic micromixers, liquid separation, optical detection, and biochemical reaction in LTCC-based microfluidic systems by Wroclaw University of Technology and Sandia National Laboratories.[32−35] Thelemann et al. found that the integrated microchannel cooling system in LTCC substrate can decrease additional temperature more than 80%, while the silver thermal vias only reduce additional temperature by about 30%.[36] With increasing heat flux, the cooling ability of a microchannel becomes much better than the metal thermal via.

The design of the LTCC microchannels should consider both the fabrication process and the mechanical reliability of the substrate. The traditional and effective cooling microchannels are straight, serpentine, and fractal-shaped microchannels. In the study of Beijing University's LTCC group,[37,38] six types of microchannel networks are supplied, including straight, serpentine, spiral, and three other fractal-shaped microchannel networks (curved, I-shaped, and parallel), as shown in Figure 3.9. The dimension of the microchannels was decided according to the theories of laminar flow and heat conductivity and the limitation of the fabrication process. The microchannel network was fabricated in the fourth and fifth layers of substrates with a cross section of 200 μm × 200 μm. A thick-film resistor of 2.0 cm × 2.0 cm as the heating source was placed on the center of the substrate surface. The experiment was controlled by the mass flow rate at the inlet and heat flux of the surface heat source.

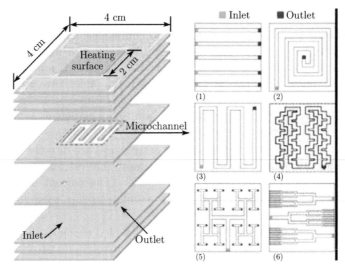

Figure 3.9 Scheme of LTCC substrate with microchannel

An X-ray was used for the observation of the internal microchannels. Figure 3.10 shows that the machining process is good enough to fabricate the microchannels accurately, even for the fractal-shaped microchannel networks. Damages such as sink, block, and distortion were found neither in the profile of the microchannel nor in the dislocation between layers.

Figure 3.10 X-Ray observation of internal microchannel and optical observation of microchannel profile.

The finite volume method (FVM) was used to calculate the temperature, fluid pressure, and flow velocity fields. A mixture of mesh was selected and refined on the sidewalls of the channels, including tetrahedral, hexahedron, pyramid-shaped, and wedge-shaped meshes. The simulation results shown in Figure 3.11 agree well with the experimental results.

The experiment results showed that the LTCC straight cooling microchannels reduce 73.4% of the temperature addition at a heat flux of 1 W/cm^2, i.e., from 79 K to 21 K, as shown in Figure 3.12. Under the same fluid mass at the inlet, the serpentine and spiral microchannels have the best ability to dissipate heat and the largest fluid pressure drop, which requires a more powerful pump. The heat dissipation ability of fractal-shaped microchannel channels is worst, which becomes the best when the pressure drop increases to the same level with the serpentine and spiral microchannels, while the temperature field of fractal-shaped microchannels is more homogeneous than other microchannels. The heat dissipation ability would also be greatly enhanced with the appearance of turbulent flow. However, more investigation should be made into the complex mechanics of fluid flow and heat dissipation of fractal-shaped microchannels before commercial application.

With the development of high-power microsystems and heat fluxes above 10 W/cm^2, 3D microchannels become more important and realizable for cooling applications in the microelectronics industry, with the advantages of small hydraulic radius, high heat conductivity, homogeneous thermal field, easy liquid control and detection, and various types of networks for different microelectronic chips.

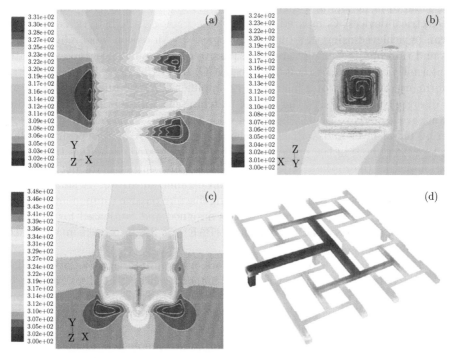

Figure 3.11   Temperature distribution at the heating surface of LTCC substrate with (a) straight, (b) spiral, (c) I-shaped microchannel networks and liquid temperature distribution in (c)

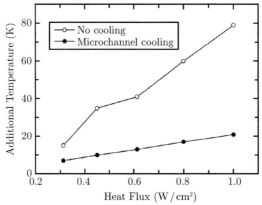

Figure 3.12   Additional temperature under cooling and no cooling condition for the straight microchannel networks

3.6.2   High-density Interconnection Substrates

HDIS are generally created on a base substrate of core material with microvia layers added to one or two sides. Smoothing the surface of a substrate to meet the requirement of the thin-film process may be the another way to improve the interconnection density on the substrate.

With the multilayer structure of LTCC, passive devices can be embedded into LTCC substrate as discussed above. However, since the minimum line width and spacing that can be achieved in LTCC is limited by the process that hinders the miniaturization of the device, thin-film technology using photolithography becomes a viable solution for the realization of small form factors and high-performance radio-frequency (RF) modules used in portable electrical devices. Compared with LTCC technology, a four to seven times size reduction can be achieved using thin-film technologies.[39]. Moreover, when photolithography processes and thin-film technology are used, passive devices on the ceramic substrate have better uniformity and tolerance than those in LTCC. ASTRI in Hong Kong demonstrated a high-performance and small form factor power amplifier (PA) module by replacing surface-mounted discrete components for the output matching network (OMN) of the PA die with thin-film integrated passive for wireless local area network LAN; 802.11 b/g) applications. A sputtering metal layer and a Plasma-Euhauced Chemical Vapor Deposition (PECVD) dielectric layer were used in the processes.[40] High-Q integrated capacitors (more than 200 at 1 GHz), inductors (more than 12 from 1 to 5 GHz), and output matching networks with high tolerance ($< 5\%$ variation) were achieved on ceramic substrate. A small form factor (with a size reduction of 67% compared with using discrete passive components) PA module with integrated OMN on ceramic substrate was designed and fabricated for wireless LAN applications. Enhanced power added efficiency (PAE) (56.3% improvement) was achieved on the PA module.

To manufacture multilayer embedded capacitors, a novel approach was presented that a variety of resin-coated copper capacitive (RC3) nanocomposite thin films ranging from 2 microns to 50 microns thick were processed on PWB substrates by liquid coating or printing processes.[41] The nanotechnology is able to produce high capacitance density in high density substrate (7–500 nF/inch^2) at 1 MHz.

Zeon Corporation and Tohoku University reported a technological breakthrough for the Cu line formation onto a quite smooth substrate surface, showing that 10 µm/10 µm pattern in line and space can be formed finely by using a semi additive method.[42] The developed insulation materials for the build-up layer package are high adhesion between the Cu line and the substrate with flat surface, extremely low water absorption of 0.2% at 100°C, low expansion coefficient, low dielectric constant of 2.6, and low dielectric loss of 0.01 at 1 GHz.

3.6.3 Flexible Substrate

A flexible substrate, such as flexible printed circuit board (fPCB), has an important role in electronic product miniaturization, since their reduced thickness allows them to bend and adapt to various shapes. For example, it has been applied in connecting two rigid substrates or functional modules of consumer handheld devices, mobile devices, and pocket electronic devices. Most flexible substrates are polyimide parylene, Poly Dimethgl Siloxane materials (PDMS), which allow for high-density via traces and tight ball pitches and are thermally advantageous. As one of the most stable organic polymers, polyimide has electric properties the following favorable of a high breakdown voltage over 1 MV/cm and a low relative dielectric constant below 3. In addition, it can be used for Foine-Pitch BGA (FBGA) interposers that closely match the die size in potable electronics. Here is an example of a parylene flexible substrate with microelectrode array for biomicrosystem application.[43]

Parylene has good biocompatibility (for C-typed parylene), an outstanding moisture barrier, excellent chemical inert ness, and extremely high dielectric strength. Additionally, parylene offers good conformability and can be easily patterned by an O_2 enrolled dry etching process. All of those characteristics make parylene superior to other materials in the applications of flexible microdevices, especially implantable microelectrode arrays (MEAs).

Recently, microelectrode arrays on a flexible substrate prepared by MEMS technology find diverse applications in neural prostheses, such as restoring hearing to the deaf, vision to the blind, and movement and sensation to the paralyzed. In all these applications, the implanted MEA works as the most important part to bridge the neurons and the outside prosthetic device for transferring the stimulating or recording charges.

A new flexible 3D MEA on parylene substrate has been developed based on the parylene transferring technique. The process can be divided into two parts: the silicon mold process (Part I) and the final parylene 3D MEA process (Part II).

Part I: The silicon mold, which is reusable for the following process, is first fabricated as illustrated in Figure 3.13. A standard 4 inch silicon wafer (Figure 3.13a) was chosen and cleaned in the piranha (H_2SO_4 and H_2O_2 solution with a ratio of 4:1 at 120°C for 10 min) before the process. A 3 μm-thick photoresist layer was spun onto the wafer and patterned as a mask (Figure 3.13b) for the deep reactive-ion etching (DRIE), in which the 3D cylinder structure with a depth of 100 μm and a diameter of 100 μm was formed (Fig. 3.13c). After photoresist removal, an isotropic wet etchant hydrofluoric, nitric, and acetic (HNA) for silicon was applied to shape the cylinders to a tip profile (Fig. 3.13d).

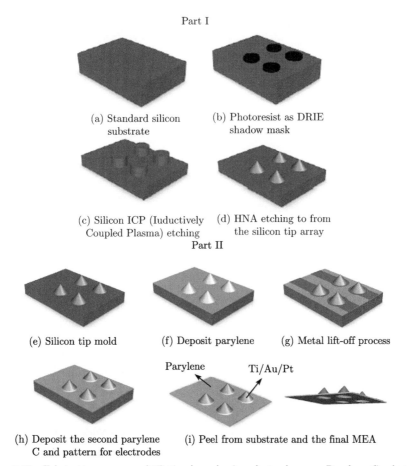

Figure 3.13   Fabrication process of 3D-tip shaped microelectrodes on a Parylene C substrate.

Part II: After the silicon mold was obtained (Fig. 3.13e), approximately 10 μm parylene C was directly deposited on the structure in a PDS2010 system (Specialty Coating System, Indianapolis, IN, USA; Fig. 3.13f). This parylene deposition process was done at room temperature, and thus hardly introduced any stress into this so-deposited thin parylene

C film and the silicon mold. The good conformity of parylene C helped it follow the 3D tip profile of the silicon substrate. A dual-layer lift-off technique (Fig. 3.13g) associated with aluminum and AZP4620 was then used to achieve the metallization on this parylene 3D substrate. About 20 nm titanium and 400 nm aluminum were adopted as the first layers to realize undercut for the second photoresist layer, 15 μm-thick AZP4620. This aluminum layer was easily etched off by the aluminum enchant (H_3PO_4:HAc:HNO_3 = 16:2:1) for about 20 min with titanium reserved. Subsequently, 20 nm titanium, 150 nm gold, and 50 nm platinum were evaporated onto the substrate. The desired metallization pattern was obtained after bathing the whole structure in the PRS 3000 stripper at 60°C for about 2 hours. After the parylene surface was roughened by oxygen plasma etching again, a second parylene C layer about 5 μm was deposited onto the substrate and selectively patterned with lithography followed by the oxygen plasma to expose the stimulating electrodes (Fig. 3.13h). Finally, after removing the left photoresist and Ti/Al, the parylene flexible MEA structure was obtained by carefully peeling off from the substrate (Fig. 3.13i). The final flexible 3D MEA on parylene substrate is shown in Figure 3.14.

Figure 3.14   A 4×5 array of 3D tip-shaped microelectrode on a parylene C substrate

Questions

(1) Why is substrate technology so important in mirosystems packaging?

(2) Analyze the pros and cons of organic PWB substrates and ceramic substrates and in what areas they are used.

(3) Summarize the characteristics of LTCC substrates, major processes of LTCC substrates fabrication, and properties of those commonly used metal materials.

References

[1] R.R Tummala. Fundamentals of Microsystems Packaging. New York: McGraw-Hill, 2001.

[2] Zhiyi Weng. Manufacturing technology of multilayer PCB with buried/blind via. Printed Circuit Information, 4 (2004): 8–9 and 20.

[3] Lixin Yao, Wuxue Zhang, and Junli Lian. The Application of AOI system in PCB. Equipment for Electronic Products Manufacturing, 5 (2004): 25–28.

[4] Mingda Shi and Xiaochun Wu. Low cost MCM and its packaging. Applications of IC, 12 (2004): 79–83.

[5] Tailun Song. Multi-Chip-Module: the Latest Assembly Technology. Microelectronics, 3 (1995): 1–4.

[6] Minbo Tian. Electronic Package Engineering. Beijing: Tsinghua University Press, 2003.

[7] Zhenyu Han, Jusheng Ma, Zhonghua Xu, et al. Progress of the research on LTCC technology for substrates. Electronic Components and Materials, 6 (2000): 31–34.

[8] Hsu J.Y., Wu N.C., and Yu S.C. Characterization of material for low-temperature sintered multilayer ceramic substrates. Journal of the American Ceramic Society, 10 (1989): 1861.

[9] Jun Wang, Hui Huang, Congxin Li et al. System of Automatic Design of Powdered Metal Forming Process. Computer Aided Engineering, 3 (2000): 20–24.

[10] Anming Li, Yejing Li, Shuhai Wang, et al. Study on the Gel-Casting Technique for 99% Alumina Substrate. Advanced Ceramics, 4 (2002): 9–13.

[11] Anbing Huang, Song Cui, and Hao Zhang. Technology for directly bonding copper on AIN. Electronic Components and Materials, 5 (1999): 31–33.

[12] Cui Song, Anbing Huang, and Hao Zhang. Study on Cofired Multiplayer AlN Substrates for MCMs. Electronic Components and Materials, 8 (2003): 25–28.

[13] Yongda Hu, Ming Jiang, Chao Bang, et al. The Study of Aluminum Nitride Co-fire Ceramic Substrate for MEMS Package. Chinese Journal of Scientific Instrument, 3 (2002): 322–323.

[14] Li, G. and Tseng, A.A. Low stress packaging of a micro-machined accelerometer. IEEE Transactions on Electronics Packaging Manufacturing, 24.1 (2001): 18–25.

[15] Xintao Zhang, Tao Tang, and Qitu Zhang. Study of Glass-mullite Ceramic Composite Substrate. Electronic Components and Materials, 4 (2004): 36–38.

[16] Zhengguo Jin, Nan Lv. Mullite Cerermics in Electricity and Optics. Contributions to Geology and Mineral Resources Research, 4 (1998): 85–89.

[17] Minhua Guo and Tingyue Wang. Soldering of SiC particulate reinforced aluminum(Al/SiC) and low temperature co fired ceramic (LTCC) materials. Materials Science and Technology, s1 (1999): 153–156.

[18] Linjun Wang, Zhijun Fang, and Minglong Zhang. Dielectric and Thermal Properties of Diamond Film/Alumina Composite. Journal of Inorganic Materials, 4 (2004): 902–906.

[19] Zhihui Wang. Thick Film Material and Process Applied in Microwave and RF Circuits. Semiconductor Information, 2 (2000): 21–29.

[20] Shuren Zhang and Yuanjun Zhang. Preparation of the high purity nanometer BeO ceramic powder. Electronic Components and Materials, 6(1999): 1–2.

[21] Zhaowen Dong. Study on the technology for LTCC ceramic substrates. Electronic Components and Materials, 5(1998): 24–28.

[22] Patricia B. Opina, Via filling of green ceramic tape, USP 4802945, 1989-02-07.

[23] T. Williams and A. Shaikh. Silver via metallization of low temperature cofired ceramic tape. International Symposium on Microelectronics (1994): 324–329.

[24] Shenli Wu. Thin-film Metalization on LTCC Substrates. Electronic Components and Materials, 6 (2001): 13–14.

[25] Zhonghua Xu, Zhiting Geng, Zhenyu Han ,et al. History and Research Focus of Low Temperature Cofired Ceramics. Materials Review, 4 (2000): 30–33.

[26] M.R. Haskard. Thick-film Hybrids: Manufacture and Design. New York: Prentice Hall, 1988.

[27] Lixi Wan, P. Markondeya Raj, Devarajan Balaraman, et al. Embedded Decoupling Capacitor Performance in High Speed Circuits, 55th Electronic Conponents and Technology Conference,

Florida, May 31–June 3, 2005.

[28] R. Tuominen, T. Waris and J. Mettovaara. IMB technology for embedded active and passive components in SiP, SiB and single IC package application, ICEP 2009, Kyoto, Japen.

[29] T. Kim, H. Kim, K. Kim, J. Jung, T. Jung, and S. Yi. Tolerance Improvement for Organic Embedded Passive RF Module Substrate, 2008 10th Electronics Packaging Technology Conference, Singapore, 1226–1230.

[30] Thelemann T. Thust H and Hintz M. Using LTCC for microsystems. Microelectronics International, 2002, 19: 19–23

[31] Neubert H. Partsch U. Fleischer D. Gruchow M. Kamusella A. and Pham T.Q., Thick film accelerometers in LTCC-technology-design optimization, fabrication, and characterization. Actuator, 2008, 1–6.

[32] Golonka LJ, Zawada T, Radojewski J, Roguszczak H, and Stefanow MS, LTCC microfluidic system. International Journal of Applied Ceramic Technology, 2006, 3: 150–156.

[33] Ostromecki G, Zawada T, and Golonka LJ. Fluidic micromixer made in LTCC technology-preliminary results. 28th International Spring Seminar on Electronics Technology, 2005, 352–357.

[34] Golonka LJ, Roguszczak H, Zawada T, Radojewski J, Grabowska I, Chudy M, Dybko A, Brzozka Z, and Stadnik D. LTCC based microfluidic system with optical detection. Sensors and Actuators B, 2005, 111–112: 396–402.

[35] Peterson, KA, Patel KD, Ho CK, Rohde SB, Nordquist CD, Walker CA, Wroblewski BD, and Okandan M. Novel microsystem applications with new techniques in low-temperature co-fired ceramics. International Journal of Applied Ceramic Technology, 2005, 19: 345–363.

[36] Thelemann T, Thust H, Bischoff G, and Kirchne T. Liquid cooled LTCC-substrates for high power applications. International Journal of Microcircuits and Electronic Packaging, 2000, 23: 209–214.

[37] Y.F. Zhang, "Microstructure, Mechanical Behavior and Microchannel Cooling for Low Temperature Co-fired Ceramic Substrate," Thesis for Ph.D. Beijing University, May of 2009.

[38] J. Zhang, Y.F. Zhang, M. Miao, et al. Simulation of fluid flow and heat transfer in microchannel cooling for LTCC electronic packages, to be published in 2009 International Conference on Electronic Packaging Technology, Beijing, China.

[39] N. Mellen, R. Chen, J. Horne, and H. Patterson. WLAN RF Module Design Tradeoffs, 2004 IEEE MTT-S IMS Workshop WSA: Wireless-LAN Solutions and RF Front-End Integration Trends, Fort Worth, June 2004.

[40] J. Liu, Lydia L.W. Leung, et al. Compact and High Efficiency Power Amplifier Module using Integrated Passive Technologies for Wireless LAN Applications. to be published.

[41] Rabindra N. Das, Steven G. Rosser, et al. Resin Coated Copper Capacitive (RC3) Nanocomposites for Multilayer Embedded Capacitors, 2008 Electronic Components and Technology Conference, 729–735.

[42] M. Sugimura, H. Imai, M. Kawasaki, et al. New Insulation Material with Flat-Surface, Low Coefficient of Thermal Expansion, Low-Dielectric-Loss for Next Generation Semiconductor Packages, 2008 Electronic Components and Technology Conference, 747–752

[43] X.J. Huang. Study on Parylene C-based 3D tip-profile microelectrode array used for retinal prosthesis, MS Dissertation, Beijing University, 2009.

CHAPTER 4

Interconnection Technology

4.1 Introduction

Interconnection technology, the fundamental and exclusive technology for microsystems packaging, is one of four basic technologies in the electronic packaging field. It is one of the research hotspots in the international microelectronics industry.[1]

Interconnection technology can be used to transfer various physical signals between chips and chips, chips and package, and devices and substrates. Interconnection of electronic signals, the main topic of this chapter, is the most fundamental technology for microsystems packaging technology. Interconnections of optical signals and of fluid signals are discussed in Chapters 2 and 7, respectively.

Braze-welding was the earliest interconnection technique used in electronic engineering. To meet the requirements for miniaturization in the microelectronics industry, three types of interconnection technologies have been developed and are being widely used. These are wire bonding (WB), tape automated bonding (TAB), and flip chip bonding (FCB).

In the microelectronic packaging field, interconnection technology has a significant influence on the performance of devices. Particularly, the interconnection of chips is crucial to the long-term reliability of electronic components. About one in every three or four malfunctions of IC devices is caused by interconnection failure of chips.

Chips typically cannot be used individually. To exchange information with the outside world, all chips are equipped with I/O interfaces. Only through the interconnection between chips and chips, chips and substrates, and function circuits and systems can the power and signals of chips be allocated and can signals be transmitted. Therefore, the fundamental functions of interconnection are to ensure that chips and devices are correctly connected to the power and ground supplies of a system and to ensure the smooth transmission of signals.

The second function of interconnection is to optimize a wiring structure of packaging. The length of physical wiring is much longer than that of power and signal wiring inside chips, and therefore how the interconnections are performed will have a significant influence on the technical characteristics of all levels of packaging systems. It is essential to define electrical parameters, such as electric resistance, inductance, and capacitance in interconnections used, so as to meet the requirements on the operation of systems.

In addition to electrical functions, some chip-level interconnections can provide mechanical support to the chips or protect the chips and reduce the stress and strain between connected materials by using adhesives. Additionally, a good electric conductor must be a good thermal conductor. Often, interconnecting wires (joints) and packaging materials also need to meet the requirements for thermal dissipation when chips are in operation. Therefore, each wire (joint) in the connection of chips acts as an electric conductor, a heat remover and a mechanical support simultaneously.

4.2 Braze-welding Technology

Braze-welding: Braze-welding is a process in which base metals are joined with some melted metals or alloys whose melting points are lower than that of the base metals. The melted metals or alloys will cool down at room temperature to form a strong, sealed joint between the base metals. The melted metals or alloys are called solders.

Wettability: The wettability of filler metal refers to the ability of the filler materials to adhere to the base metals. The stronger the adhesion is, the better the wettability. This adhesive force can be defined as the contact angle θ between the solid metal and the filler metal shown in Figure 4.1 and can be calculated through the Equation 4.1:[1]

$$P_{\mathrm{SF}} = P_{\mathrm{LS}} + P_{\mathrm{LF}} \cos \theta \tag{4.1}$$

In this equation, P_{SF} is the surface tension between the filler metal and the flux, P_{LS} is the surface tension between the melted metal and the solid metal, and P_{LF} is the surface tension between the filler metal and the flux.

When the angle θ is 180°, it means that the filler metal is totally non wettable, and when the angle is 0, it means the filler metal is fully wettable. When the angle is smaller than 90°, it means the filler metal has good wettability. For instance, the wetting angle of the ^{63}Sn/^{37}Pb eutectic solder on the clean surface of copper is about 20°, indicating that the filler metal has good wettability.

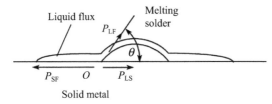

Figure 4.1   Wettability of solders

Flux for braze-welding: Most base metals are oxidized in the air, forming a 2–10/nm oxide film on their surface. This film causes an inferior wettability of the metal welded and thus deteriorates the quality of braze-welding. Fluxes are used to remove the oxide film on the surface of base metals or filler metal at the braze-welding temperature to improve the wettability. Colophony, which we use in welding devices, resistors, and capacitors in assembling a radio sets, is a typical flux. The following factors should be taken into account when choosing fluxes.

(1) They should have strong enough chemical activity.

(2) They should have good wettability and fluidity.

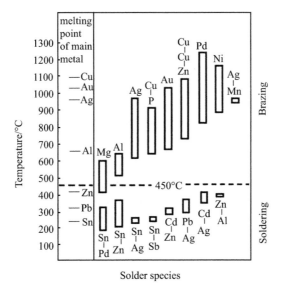

Figure 4.2   Common filler metals and ranges of their melting points

(3) They should have good thermal stability.

(4) They should have certain electrochemical properties.

(5) They should be harmless to human, microsystems and the environment.

(6) They should be easy to remove.

Categories: Based on their melting points, filler metals can be divided into two categories: brazing filler metals and soldering filler metals. The melting points of brazing filler metals are higher than 450°C, while the melting points of soldering filler metals are lower than 450°C. Correspondingly, welding can be categorized into brazing and soldering. Figure 4.2[1] is a list of common filler metals.

The eutectic solder ^{63}Sn/^{37}Pb, with a melting point of 183°C, has been the most widely used solder in electronic engineering interconnection for decades. In order to meet requirements for environmental protection, researchers are developing and using lead-free solder series such as Sn-Ag, Sn-Zn, and Sn-Bi.

4.3 Wire-bonding Technology

4.3.1 Basic Concept[2]

Wire-bonding is a method whereby a chip with active surface facing up is attached to a package or a substrate and then the I/O pads on the chip are connected with the I/O leads of the package or the wiring pads on the substrate with a fine wire such as gold or aluminum wire.

Wire-bonding plays a very important role in interconnection technology in electronic engineering for its maturity, simple technique, low cost, and strong applicability, and it is widely used in microsystems packaging. It is still in change and developing to cater to and satisfy the demands of new techniques and materials of semiconductors.

For over ten years people have been predicting that it will be outdated in the near future. This technology, however, didn't disappear; it is still actively used from low-end to high-end packages as one of the mainstream interconnection technologies, and continues its development along with the development of microelectronics systems technology.

4.3.2 Types of Wire-bonding[3,4]

Depending on the degree of bonding automation, wire bonding can be divided into manual bonding, semi/automatic bonding, and auto bonding. Based on the bonding process, it can also be divided into three types: ultrasonic bonding, thermocompression bonding, and thermosonic bonding. These three processes have their own features and are applicable for different products.

Aluminum wire bonding is usually performed in an ultrasonic bonding process at room temperature. In this process, the high-frequency vibration energy generated from an ultrasonic generator is transferred to the contact surfaces of the aluminum wire and bonding pad through the bonding head under a compression force. A firm bond is formed by the "atomic bonding" between the two intimate contact surfaces. The bonding head is typically a wedge type. Therefore, this kind of bonding is also called wedge bonding.

Thermocompression bonding involves the use of temperature and pressure to produce plastic deformation of metal on the bonding pad, as well as destroying the oxide layer on the interface of the metal bonding pad so as to make the interface of both bonded metal wire and bonded pad metal reach atomic attraction bond and realize bonding through the attraction force between atoms. In addition, the rough metal interface together with the use of temperature and pressure can help close embedding between the metals. This bonding process, however, may cause damage to metal wire from over-distortion, as well as affecting bond quality. This limitation restrains the use of thermocompression bonding.

Thermosonic bonding is also called gold ball bonding. The principle of thermosonic bonding is similar to that of thermocompression bonding, except that themosonic bonding also applies ultrasonic energy. The process of thermosonic bonding is shown in Figure 4.3: (1) melt one end of metal wire into a ball through high-voltage electric flame off; (2) apply temperature and ultrasonic pressure energy on the bonding pad of the chip to produce interface plastic deformation and destroy the oxide film of the interfaces for activation; (3) make diffusion between the two interfaces to be bonded through contacting two metals to form the first bonding spot; (4) move the bonding head to the bonding point of the leadframe or the bonding pad of the substrate through precise and complicated dimensional control; (5) apply temperature and ultrasonic pressure for the bonding of second bonding spot; (6) raise the controlled tail length of the bonding wire for the next ball formation; (7) repeat (1–6) steps and finish the connection of the second wire, the third wire, etc. The obvious difference between the two bonding joints is, during the processing of the first bonding joint, the end of the metal wire should be melted to a ball shape, while in the processing of the second bonding joint, it is not necessary to melt the end of the metal wire into a ball shape; the pressure from the specific shape of the wedge snaps the metal wire instead of melting it. Since thermosonic bonding occurs at a temperature lower than that for thermocompression and the bonding strength is enhanced, it has become the main bonding method in Au wire bonding. More than 90% of the bonders used in production lines today are automatic Au wire bonders adopting the thermosonic bonding technique.

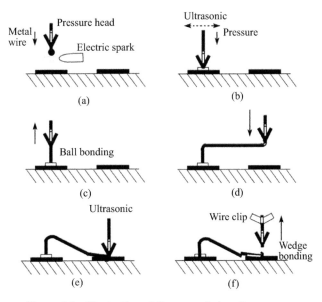

Figure 4.3   Illustration of thermosonic bonding process

4.3.3   Materials Used in Wire-bonding

Different bonding materials will be chosen by different bonding methods. A gold wire, for example, is mainly used for thermocompression bonding and thermosonic bonding; an aluminum wire or aluminum alloy (Si-Al, Cu-Si-Al) wire is usually used for ultrasonic bonding. Basic requirements of materials for wire bonding are as follows:

(1) They are capable of forming low-resistance ohm contact with bonding material, such as aluminum, gold, or other interface materials.

(2) They have high bonding strength with bonding materials.

(3) They have good conductivity.

(4) They have good plasticity for bonding technique and good shape stability.

(5) They are chemically stable.

Tables 4.1 and 4.2 list the features of gold wire and aluminum wire, respectively.[2,4] They show that both gold wire and aluminum wire have greatly improved ductibility and flexibility after the annealing process and they become more suitable for defect-free bonding. Since there is a different melting point between aluminum and gold wires, the annealing temperature must be different to achieve the ideal effect. For example, the annealing temperature for aluminum is relatively low, while that for gold is slightly higher.

Table 4.1   Properties of gold wire used for wire bonding

Diameter (μm)	Weight (mg/m)	Average minimal ductibility		Failure strength(N)	Resistivity (Ω/m)
		Before annealing(%)	After annealing(%)		
127	244	5	15	1.6	1.84
76	88	5	15	1.05	5.32
50	39	5	15	0.26	11.96
25	9.78	5	6	0.06	47.6
12.5	2.49	3	4	0.015	192
5	0.39	—	2	0.0024	776

Table 4.2   Properties of aluminum wire used for wire bonding

Diameter (μm)	Average minimal ductibility		Failure strength(N)		Resistivity (Ω/m)
	Before annealing(%)	After annealing(%)	Before annealing	After annealing	
127	1.5	15	4	2	2.23
76	1.5	15	1.4	0.7	6.12
50	1.3	12	0.65	0.32	13.9
25	1.2	4	0.15	0.08	55.7
12.5	1	3	0.04	0.02	223
25 (1% Si)	1	3	0.18	0.028	—

In addition to gold and aluminum wire, copper has also been used for IC interconnection in recent years. Since the gate delay of an integrated circuit depends on the resistance and capacitance of interconnection material, copper with a higher conductivity than that of conventional aluminum is good for reducing resistance and gate delay. The use of copper as an interconnection material, however, depends on improving the bonding technique. The protection layer of a copper bond pad, for example, is required to prevent the oxidization of copper. During wire bonding, it is necessary to use the multilevel drive of the ultrasonic energy converter, which can destroy the oxide layer on the surface of copper with high frequency ultrasoni cenergy and finish diffusion bonding with lower frequency ultrasonic energy.

In addition to its advantage of low resistance, copper wire has lower cost and higher strength and stiffness in comparison with gold, and it is suitable for fine-pitch wire bonding. Copper also has low intermetal diffusivity, and its intermetallic compound grows slowly so that the electric resistance of the infiltrating layer between metal layers is low. For copper wire bonding, protective gas should be used in electronic ignition to prevent copper ball oxidation.

4.3.4   Key Process Parameters of Wire-bonding[4,5]

(1) Temperature control: The temperature range is from room temperature to about 400°C. Temperature control accuracy is better than 1°C. Temperature can be controlled manually or automatically by controlling the current of the heater.

(2) Precisely positioning control: Precisely positioning control of the chip, lead frame, and package substrate is usually accomplished by a combination of precisely controlling the guide track, precision jigs, and vision system.

(3) Setting of bonding parameters: parameters[6], such as current, voltage, frequency, and amplitude for driving the ultrarsonic energy converter, wire clip, electronic flame, bonding pressure, and time are carefully determined to ensure the precision of bonding spot, bonding quality, and long-term reliability.

Figure 4.4 is a sketch map of the bond head of a bonder. The manufacturing of the ultrasonic energy converter, wire clip, and wedge requires advanced fabrication techniques and inspection/testing skills and thus is more valueable. The key technology involved in this part includes: program-controlled ultrasonic energy converting technique, gold ball forming technique, bonding pressure control, and wire clip precision movement control.

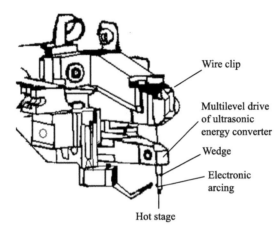

Figure 4.4   The structure of a bond head

The ultrasonic energy converting technique involves precision supporting wedge, the vibration energy, and pressure transferred to the wedge. The key point of the technology is the control of output frequency and voltage of the ultrasonic generator to regulate the resonance frequency and amplitude of the ultrasonic energy converter, which is very important to the quality and reliability of the bonding spot.

The way a gold ball is formed has a direct effect on the performance of the ball bonding joint. It controls the time and energy of the high voltage electric charge to melt the tip of the gold wire and form a small ball under the surface tension. In gold wire bonding, in order to control the bonding quality, consistent ball diameter is required. Thus, automated bonders are usually equipped with ball diameter detectors. Gold ball diameters are typically controlled within 1.4 to 3 times the diameter of gold wire.

Bonding pressure is precisely controlled by a DC motor driving through the feedback signals of a proximity sensor. The bonding force is typically controlled within 10–20 grams with a resolution of 0.002 Pa. The wire clip driver is driven by a piezoelectric ceramic actuator. Its performance in opening and closing is very important to ensure a rapid and accurate bonding process. When the wire clip is closed, enough holding force should be used to ensure the snap of wire without damaging the integity of wire. When the wire clip is opened, enough space should be provided to let the wire through without any obstacles.

Though it is easy to achieve single drive and control of every functional part of the bonding sets, it is not easy to fulfill these performances rapidly, precisely, orderly, and harmoniously, because they are an integrated unit assembled together with dependent and cooperative functions. It is a remarkable indication of qualified assembling testing, a symbol of reasonable electric design and software control, as well as the key point of reliability and joint bonding precision. Therefore it is necessary to draw a precise sequence diagraph showing the electronic flame, ultrasonic energy converter, clip drive, bonding pressure, and vertical movement of wedge in order to fully understand the rules based on technology support and to provide logistic support for electro-mechanic design and software control.

4.3.5 Technical Limitations[5]

Wire bonding technology has some technical limitations, including (1) the parallel connection of multiwire may cause adjacency effects resulting in the uneven distribution of the current between bonding wires on the same silicon chip or bonding wires on different chip of the same module; (2) the large parasitic inductance of the bonding wire may cause high switch overvoltage; (3) the long and fine wire for planar packaging structure is not effective for thermal management.

4.4 Tape Automated Bonding Technology[6,7]

4.4.1 Basic Concept

Tape automated bonding (TAB) is an important packaging technology assembling microsystem chips onto a metallized flexible polymer tape, which is a polyimide metallized film with tractor feed fixturing holes on two sides, and it works to support the chip and the interconnection wire between the chip and the surrounding circuits.

The TAB technique starts with the formation of a ball bump on the chip and preparation of flexible and thin organic tape with inner and outer I/O counts; then the balls on the chip and the inner counts on the tape are bonded together through pressure bonding, and encapsulant is further dispensed on the chip for protection; after that, the outer counts and the bonding pads of the wiring pattern on the substrate will be interconnected through thermocompression or braze-welding. The main techniques are summarized as follows: tape fabrication, bump formation, wire pressure bonding, and encapsulating technology.

The TAB package is used for automatic assembly and mass production, and it can also meet the requirements of high-density I/O and high-speed and large-scale integrated circuit packages. This technology was developed for overcoming the disadvantage of wire bonding in producing smaller and higher–density I/O devices in the electronic package field in the 1960s. Until the 1980s, TAB had been thought of as the main direction for developing chip's high-density package. The disadvantage of TAB is its high cost, making it only suitable for mass production. Therefore, since the 1990s, with the rapid development of BGA and CSP as well as flip-chip (FC) bonding technology, the application of TAB technology has been restrained a lot in recent years.

4.4.2 Essential Techniques[8−10]

1. Tape Fabrication

The various tapes for TAB packaging can be defined according to different standards of structure and application. There are single-layer metal tape, double-layer (polyimide and copper), and triple-layer (polyimide, adhesive layer, and copper) tapes. The single-layer metal tape has the advantage of low material cost, simple manufacturing process, and good thermal resistance, but the electrical performance of the chips cannot be tested after inner wire bonding. Double-layer and triple-layer tapes can be used for high-density circuits for mass production, in which the chips can be tested electrically after inner wire bonding. Double metal layer tape can improve signal characteristics and is used for high-frequency devices as well.

The tape is usually manufactured by a photolithography process etching the copper foil. The thickness of the foil is determined by the precision of the circuit and the strength of the wire as required. The equipment for manufacturing tapes is quite complicated and expensive. Standard commercial tapes from professional companies for integrated circuit packages can be ordered. These tapes can be manufactured into a long tape, which, like a filmstrip, can be rolled for the convenience of automated production.

A good bonding layer with gold or tin coating is required for a tape bonding pad with a thickness about 1 μm. Matching the chip's pad with the tape's wire in chip and tape design should be carefully considered. The width of the tape wire and the spacing between the centerlines of adjacent wires should vary with different products. A typical width of the tape wire is about 50 μm, while the spacing between the centerlines is 100 μm.

2. Bump Formation[11,12]

The bumps on the chip connecting to the tapes act as a kind of electrode or pad for the chip. Usually, a thick metal layer will be added onto the bond pad of the chip as a bonding area. The typical bump structure is shown in Figure 4.5. The structure of bump includes adhesive layer, barrier layer, and bonding metallic layer, usually in the structure of titanium/tungsten/gold.

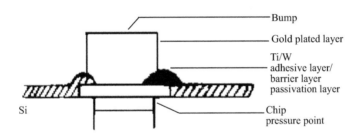

Figure 4.5   Schematic bump structure

The rule in designing bumps is that the diameter of the passivation hole should be smaller than that of the bonding pad area, while the size of the bump should be bigger than that of the passivation hole and smaller than the area of the bonding pad metal. There are two advantages for this design: first, the bonding pad metal can be completely covered by bump metal, which is good for avoiding erosion; second, damage can be avoided around the bonding pad during pressure bonding.

Bumps are mainly mushroom shaped or column shaped. Gold plating, which can help create the smallest contact resistance and form uniform bumps, is a key technique for the bumping process. The height of a gold-plated bump is usually between 20 and 30 μm. Bump coplanarity is much more critical in gang bonding than that in single-joint bonding. Generally, bump coplanarity in one chip should be within ±5%, and on the same wafer within ±10%.

3. Wire Pressure Bonding[13]

Wire pressure bonding can be divided into inner wire pressure bonding and outer wire pressure bonding. The former is to bond inner bond spots on the tape and bumps on the chip together. Mature automated production usually adopts a multispot bonding or batch bonding technique, that is, finish the bonding of all bond spots on one chip in one step. Single spot bonding is usually used in earlier stages of a product's R&D[14].

A reflow or thermocompression bonding can be applied in bonding inner wires. The reflow process requires all metal structures to be melted at the same time, and to form an alloy with other metals. Au-Sn structure, for example, can be created with reflow bonding.

The pressure bonding of outer wire, similar to that of inner wire, also has three bonding methods, reflow, thermocompression, and thermosonic, for two processes of gang bonding and single-joint bonding. Outer wire pressure bonding usually should be done after encapsulation and testing.

4. Encapsulation

During chip encapsulation liquid encapsulant is dispensed onto the active face of the chip and cures it, providing the chip protection from moisture. This process is performed after inner wire pressure bonding.

Properties of encapsulant materials to be considered are as follows: solvent resistance, curing procedure, preparation method, mechanical properties, thermal expansion coefficient, ionic contamination level, and final working environment. Epoxy resin, silica gel, polyimide, etc. are common encapsulant materials.

Epoxy resin is the most commonly used encapsulant; it provides ease of operation, similar thermal expansion coefficient to silica gel, low cost, and simple technique, i.e., easy to cure at room temperature or above. It can be applied with an automated dispenser in mass production. Despite its high cost, polyimide is an ideal encapsulant with excellent performance, better than that of epoxy resins.

The encapsulated chips need to be tested, burnt in, and sorted before applying outer wire pressure bonding.

4.4.3   Technical Features

A TAB packaged device has smaller size, lighter weight, and thinner shape. Other features are as follows:

(1) It has good flexibility.

(2) It has high-density input and output counts.

(3) It has excellent electric performance, applicable for high-frequency circuits.

(4) It has high efficiency in thermal management.

(5) It is applicable for automatic assembling. Particularly, inspection and testing can be done automatically to detect defects in early stages.

4.5   Flip-chip Bonding Technology

4.5.1   Basic Concept[13−16]

Flip-chip bonding (FCB) is a process in which bumps are fabricated on a bare chip's electrode and then interconnected with packaged substrate by using soldering or other techniques with the chip's active face down. This technology replaces interconnection wires with bumps, which are used in traditional wire bonding. The area array I/Os can fit for large-scale and ultra-large-scale integration circuits with the highest assembly density.

FCB was developed in the 1950s–1960s and is now used widely in aviation and avionics, communication, computer science, the automobile industry, and so on. Flipping a chip for interconnection has brought about a series of reformations in microelectronic packaging and has led the direction in high-density packaging. Typical examples include wide application of BGA, rapid development of CSP, and extensive use of C4 (controlled collapse chip connection) in high-frequency MCM.

Flip-chip bonding has many advantages, such as high precision, small size of hybrid integrated chip, high density of I/O, short interconnection wires, and small parasitic parameters. In flip-chip bonding using reflow soldering, the self-aligned effect can be achieved by using the surface tension of melted filler metals, and thus high precision passive alignment can be realized. Therefore, flip-chip bonding is the first choice for advanced integration of high-density chip/chip interconnection.

In addition, in order to achieve high mechanical strength and reliability of devices, the space between chip and substrate can be filled with encapsulants with high thermal conductivities. In the past few years, the flip-chip package has gained increasing popularity as the preferred solution for high performance Application Specific Integrated Circuit (ASIC) and microprocessor devices, with a trend toward more I/O, finer pitch, and larger chip size. In addition, organics substrate has been employed instead of the more expensive ceramic substrate for the cost consideration.[29]

4.5.2 Bonding Process

Flip-chip bonding is composed of bump preparation and flip-chip assembling.

The first step of bump preparation is to prepare the under bump metallization (UBM) on the contact pads of a chip. The functions of UBM are to protect the electrodes on the chip from corrosion and to avoid any other failures, such as cracking and breakage of bump or electrode, caused by incompatibility between bump material and electrode material. The material of the UBM layer is selected based on the materials of the electrode and the bump and the bonding temperature. When the electrode is made of aluminum using the silicon-based Complementary Metal-Oxide-Semiconductor (CMOS) technique, and the bump is Pb-Sn or lead-free, then materials such as Sn-Cu-Ag, Ni, Au, and TiW alloy are used for UBM,[16] as is shown in Figure 4.6.

Bumping is the most complicated step of flip-chip bonding. There are different bumping methods for different bump material. Based on the melting temperature, all the bump materials can be categorized into low—melting point materials, such as lead, tin, indium, and their alloys, and high melting point materials, such as aurum and platinum. Low melting point materials, which are widely used in bump fabrication, have advantages such as low soldering temperature, good plasticity, and CTEs of the two base metals to be bonded that are easy to match. A few examples of bumping methods will be presented in this section.

The combination of photolithography and electroplating or any other metallic film formation technique is the most common method, which can form a bump array with high precision, good coherency, tiny spacing, and high density. The method of combining screen printing and reflow soldering can well satisfy the needs for cost reduction, manufacturing process simplification, and large-scale production. In this method, solder paste is placed on a passivated electrode pad via screen printing and then heated to melting temperature. Since the wettability of the pad is better than that of the material around the pad, solder can be converged onto the pad, forming a bump in the ball-coronal shape by surface tension. The size and height of the bump depends on the size of the pad and the amount of the solder paste.

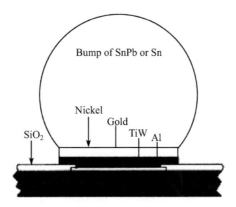

Figure 4.6 The structure of the bump and passivation layer in flip-chip bonding

This method greatly saves cost without the use of expensive facilities, such as photolithography and electroplating equipment. However, we cannot precisely control the amount of solder pastes, and as a result this method cannot be used in making bump arrays with high precision.

In the flip-chip assembling process, the bumps on chip are bonded to the corresponding metal pads of substrate using a flip-chip bonder. The bumps should be covered with flux before being aligned to the pads. After the alignment, the bumps with a low melting point are reflowed at high temperature and then form solid electrical and mechanical connections at room temperature.

To enhance the reliability of bonding joints, the space between chip and substrate is filled with resin. Typical FOB steps are illustrated in Figure 4.7.

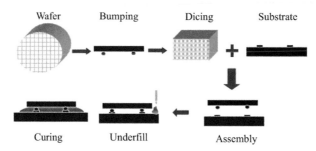

Figure 4.7   Typical steps of flip-chip bonding

4.5.3   Main Materials of Flip-chip Bonding[17,18]

1. Chip and Substrate

The interconnection between flip-chip circuits and outside is realized through soldering balls. Generally, aluminum is used in circuit wiring on most of silicon chips. Since aluminum is metallically active, UBM is required. To improve the reliability of solder joints, the UBMs for flip-chip are multilayer metals, which consist of three layers:
  (1) Adhesion layer, typical materials are Cr, Ti, V, and W.
  (2) Barrier layer, typical materials are Ni, Cu, Pd, and Pt.
  (3) Wetting layer, typical materials are gold films, silver films, or gold alloy films.

2. Flux

Flux performs the following functions: (1) Chemically prevents oxidation during the soldering process; (2) Thermally raises the temperature of base metals to be bonded high enough to wet filler metals; (3) Physically improves the wettability of filler metals and covers the entire soldering surface with filler metals.

Additionally, since there's no solder paste on the substrate, flux will perform the function of the solder paste to adhere to the chip. The amount of flux plays an important role in flip-chip bonding. Too little flux cannot remove oxide on the surface of bumps and clean substrate, while with too much flux, the chip will slide on the substrate and float on flux during reflow soldering, thus producing defects in interconnection. Also, when the bump pitch decreases, the large surface tension of melted solder will attract neighboring bumps, leading to a short circuit. Moreover, too much flux will cause difficulty in back-end cleaning.

3. Underfill

Underfill can reduce thermal mismatching between chip and substrate effectively, improve the reliability of bonding joints, and extend the thermal fatigue life. The lifespan of solder joints is highly dependent on the parameters of underfill.

4.5.4   Reliability of Flip-chip Bonding[3,18]

Many factors could affect the reliability of flip-chip bonding-material, structure, process, etc. Worldwide, the study on the reliability of flip-chip bonding is mainly focused on the following areas.

1. Reliability of Solder Joints[17,19−24]

(1) Shape and height of solder joints and solder composition.

The shapes of solder joint are determined by the wetted area of the interfaces of a chip and substrate, the volume of solder, and the weight of the chip. Generally, a "truncated ball" shaped joint will appear between the bonding interfaces of the chip and the substrate, and the joint height depends on the radius of the wetted area, the volume of the solder, the total number of joints, and the weight of the chip. The height of solder joint refers to the space between the chip and the substrate and is one important factor affecting the reliability in thermal cycling. It has been shown that the fatigue life of a solder joint increases with the increase in the height of the joint. Using the method of stacking solder joints can generate high joints and therefore increase the service lives of solder joints.

The composition of solder ball material has a big influence on the reliability of a solder joint as well. Compared with a Pb-Sn based solder joint, a Pb-In solder joint possesses a better resistance to fatigue. Thermal cycling tests show that adding 15%–20% indium can significantly extend the lifespan of a solder joint. Pure indium solder joints in optoelectronic devices have good reliability.

(2) Analysis of stress and strain of solder joints.

The analysis of the stress and strain of flip-chip solder joints is the basis of the study on the reliability of the solder joints. In the thermal cycling test, strain range results in the accumulation of plastic strain and leads to the failure of solder joints, as a result of the alternating action of creep and fatigue. The analysis of stress and strain of flip-chip solder joints consists of two aspects—the analysis of stress and strain of solder joints in systems with and without underfill.

In a flip-chip bonding system without underfill, the dominant deformation of solder joints is shear deformation, a result of the horizontal displacement of the chip caused by the mismatching of the CTEs between the chip and substrate. The thermal fatigue reliability of solder joints is closely related to their shear strain. On the whole, the shear strain of a solder joint and the distance from the joint to the neutral point (DNP) are in a linear relationship. Therefore, when the size of a chip increases, the number of solder joints distributed on the edge of the chip will increase, resulting in an increase in the DNP of solder joints. This would affect the thermal fatigue reliability of the system.

Filling underfill in the space between a chip and substrate can relax the CTE mismatching and significantly enhance the thermal fatigue reliability of solder joints.

(3) Life prediction of solder joints.

Life prediction has always been a main topic in thermal fatigue reliability of solder joints. Strain range and strain energy density are the main failure parameters used in life prediction of solder joints. They can be used to estimate the thermal fatigue life of solder joints.

The failure of solder joints in thermal cycling can be attributed to the initiation and propagation of thermal fatigue cracks. The cumulative damage model–based method of crack initiation and propagation is an important means for assessment of the thermal fatigue reliability of solder joints. There are currently two life prediction methods based on the cumulative damage model. One is a fracture mechanics–based crack growth model, and the other is a cumulative damage–based energy model. The results based on the two models are consistent with experimental results and, therefore, both methods have found applications in the study of thermal fatigue reliability of solder joints.

2. Material and process of underfill, and reliability

The basic principle of underfill is shown in Figure 4.8. Without underfill, if the CTE of the substrate is greater than that of the chip, the expansion of the chip is smaller than that of the substrate in heating (see Fig. 4.8a), while in cooling, the chip will experience a shrinkage smaller than that of the substrate (see Fig. 4.8b). When uderfill is applied, the thermal deformation of the substrate is locked to that of the chip and both would be almost the same (see Fig. 4.8c), thereby reducing the shear deformation of solder joints.

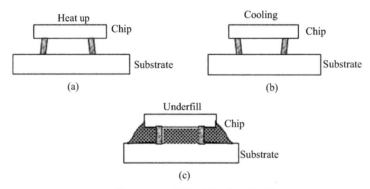

Figure 4.8   Principle of underfill

Therefore, the thermal fatigue life of flip-chip solder joints depends, to a large extend, on the mechanical properties of the underfill, such as CTE, Yang's Modulus (E), bonding strength, glass transition temperature (Tg), etc. Of them, the selection of CTE and E is the most important. Studies have shown that when the CTE of the underfill matches the solder joint connecting the chip and the substrate, the stress caused by the thermal mismatching of the chip and the substrate can be reduced effectively, and the thermal fatigue life of solder joints can be improved. When the CTE value of the underfill is slightly lower than that of the solder joint, the equivalent plastic strain range of the joints reaches a minimum value, and it will also decrease in line with the increase of Yang's modulus E.

The CTE value of underfill is adjusted through the change in the amount of fillers and the chemical properties of polymers. The change of the maximum stress of solder joints has little to do with the amount of fillers. Therefore, the amount of fillers has little impact on the reliability of solder joints. However, amounts of fillers that are too small will lead to an increase in the CTE value of underfill, while amounts of fillers that are too large will lead to a drop in the fluidity of underfill, thereby affecting the underfilling process. Also, poor fluidity of underfill could lead to incomplete filling, which will leave cavities in the filling layer and then induce reliability issues. If cavities occur near a solder joint, a local stress concentration may take place and lead to an early failure of the joint.

Underfill needs to be cured after filling. During the curing process, the contraction of underfill and the mismatching of CTEs during cooling will produce a high stress on the bonded interface of the chip. In some cases, the stress is high enough to cause chip cracking.

In addition, a fillet around the edge of the chip will be formed after curing. The fillet may have certain effects on the reliability of solder joints. Both too large and too small filler will produce high strain in solder joints. Improving the shape of such a fillet could reduce the stress on solder joints. Therefore, the selection of underfill material, underfilling process, and the shape of the fillet are all very important factors affecting the quality of flip-chip bonding.

4.6   Chip Interconnection in System-level Packaging

4.6.1   Chip Interconnection in MCM

The leap in development in electronics technology has led to integrated circuits that are increasingly sophisticated, and most functions of an electronic system can now be performed by one packaged chip or an integrated module, marking a great shift in microelectronic packaging from device level to integrated system level. Correspondingly, the relevant interconnection technology should also keep up with the pace of the change. The traditional method used in system integration is to mount several types of ICs or chips and passive components on one PCB substrate using interconnecting wires. Unlike the traditional method, the MCM technology integrates multiple chips in one packaged system, which is independent of other systems. In general, the MCM technology is characterized by high packaging density, small size, light weight, short time of signal transmission, high reliability, and maybe low cost.

The interconnection of chips is one of the most important components of the MCM technology. Wire bonding, tape automated bonding, and flip-chip bonding are the core of the MCM interconnection technology. Typical applications in MCM are as follows:

The challenge of interconnection in 3D MCM is to achieve 3D high-density packaging. 3D MCM refers to the method of arranging electronic components vertically or in the z direction as well as horizontally or on the x-y plane, namely, packaging all chips and components in multiple layers in one carrier. There are three major 3D MCM methods. (1) To stack chips and interconnect them by using the wire bonding before connecting them to substrates. (2) To stack 2D MCMs, where every 2D MCM performs one or more functions, and assemble them to form a subsystem or a microsystem. (3) A combination of the previous two methods. Chips are stack and bonded onto substrates and then the substrates are stacked up by wire bonding interconnection. This is a packaging method superior to the other two.

Dozens of 3D MCM methods are being widely used by a variety of companies. For example, Texas Instruments used Si-substrate TAB vertical interconnection technology to develop and produce ultra-high-density memory units. Panasonic presented a wire stacking bonding method for interconnecting stacked chips, which connects chips vertically using the TAB technique. The Grumman Space Technology Co., USA, applied a flip-chip bonding onto the side faces of stacked chips to produce military-use monitoring equipment.[25,26]

It is very difficult to integrate MEMS and IC in one chip to form a monolithic integration. MCM packaging supports the assembly integration of many types of chips, and only little modification is required to MEMS or microelectronic manufacturing technology. Therefore, the packaging method is an effective solution to integrate MEMS chips with IC chips. It uses high-density interconnection (HDI) technology, with bare chips embedded in the substrate in some cases.

MCM-D is a traditional HDI method, whose process is first depositing the interconnection layer on the substrate and then installing chips on the interconnection layer. Wire bonding is the main interconnection method used in this process.

Some MEMS chips need special physical paths to communicate light signals and fluidic media with the outside. It is possible to modify existing HDI technology for fabrication of MEMS devices' access channels. Figure 4.9a is the process of HDI, and Figure 4.9b is a modified HDI process, in which laser ablation technique is added to produce the I/O interfaces for MEMS devices. The laser ablation technique can selectively etch out a window in the insulation layer on the top of the chip.

Ceramic multichip module (MCM-C) assembles various chips on an AlO-based multilayer substrate. MCM-V stands for the vertical multichip module,[28] with lead wires and through-holes printed in different layers. All layers are cofired under sintering temperature simultaneously. The metal parts, such as the lead frame and cooling fins, can be bonded using the eutectic process.

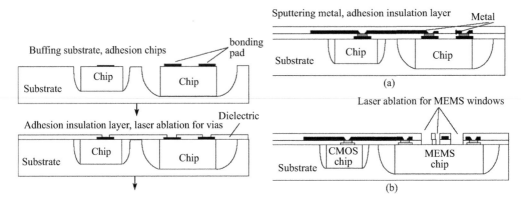

Figure 4.9   HDI technological process

4.6.2   Interconnection in SIP[27]

System in package (SIP) is a new-generation microsystem integration technique, in which functional bare chips and individual components are integrated in one or stacked substrate to function as one system.

Chip-level interconnection is also the core technology used to realize high-performance SIP. Wire bonding, tape automated bonding, flip-chip bonding, and any other interconnection technology that can directly connect IC chips can be used to achieve the high performance of SIP.

Worldwide, many large electronics companies and research institutions are developing SIP technologies and relevant products. For example, Fujitsu developed a new SIP technique, CS module (chip-scale module), in 2002. Compared with old SIP techniques, the new technique greatly reduces packaging area. In this technique, chips are stacked on bare chips using protective film. Copper wiring interconnection other than wire bonding technique is used between chips. Most importantly, all assembly processes of CS module are conducted in the wafer level. The precision of bonding chips onto wafers is ± 5 μm.[3]

4.7   Advanced Interconnection

With the development of SOC technologies, high-density interconnection is required. Thinner modules based on advanced substrate drive the need for less than 30 μm pitch interconnections. The computer and communication technologies with multicore processors for the highest aggregate systems are pushing the I/O density to more than $10,000/cm^2$ and pitch to less than 50 μm. In addition to the progress on studying fine pitch BGA, FC, and microbump, advanced interconnection for next generation electronics focusing on nano-interconnection, 3D interconnection has been reported.[30−33] Furthermore, the Iuternatianal Technology Roadmap for Semicondueton (ITRS) has predicted that the bump pitch of peripheral flip-chip will be reduced to 20 μm by 2009.[34]

4.7.1   Nanointerconnection

Carbon nanotube (CNT) based nanointerconnection technology has been a research interest for several years. CNT, if tamed, could be the copper replacement, with the growing awareness that copper is unsuited to future Large Scale Iutegrateod Circuits (LSI) dimensions. Fujitsu reported setting a packing density record of 9×10^{11} CNT/cm^2, and creating CNT vias of 160 nm in diameter.[35] The Rensselaer Nanotechnology Center presented a cooling chip through aligned "fins" of CNT. L. Jia et al. from Shanghai Jiao Tong University demonstrated a new approach for ultra-fine pitch chip on glass (COG) bonding, called particle on bump (POB) technology.[36]

COG technology, based on anisotropic conductive films (ACF) for fine pitch flip-chip interconnection, is widely used in flat panel display (FPD) modules for connecting driver ICs to the displays, especially for small size panels, which will be discussed in Chapter 7. An ACF usually consists of an adhesive epoxy matrix with conductive particles dispersed in it. The conductive particles are metal-coated polymer spheres 3–5 μm in diameter, and the adhesives are thermosetting resins. During the COG bonding process, the conductive particles trapped between the bumps of driver IC and the ITO (Indium Tin Oxide) electrodes of the module panel establish the electrical connections.

As the demand increases for higher resolution and cost reduction, the bump pitch of the driver ICs becomes finer and finer. The current ACF-based COG technology is confronted with two major issues. One is the contact resistance. As the cross-sectional area of the bump decreases with the pitch reduction, the number of the conductive particles trapped between the bumps and the ITO electrodes would decrease. The possibility of having too few particles to fulfill the contact resistance requirement or even forming electrically open joints increases. The other is the short. As the gap between the bumps becomes smaller, the chance of forming electrical shorts due to agglomeration of conductive particles between the adjacent bumps increases. It was demonstrated that the POB process can be implemented in the wafer bumping process. The interconnections formed by POB technique overcome the intrinsic problem with ACF, i.e., conflicting requirements for a large bump contact area to capture enough particles and a large bump gap to avoid shorts with increasingly finer pitch. In the POB technique, the bump size and bump gap are no longer dominated by the conductive particles. Therefore, we can conclude that the POB process is a low-cost viable technique, which has the potential to replace ACF to provide ultra–fine pitch flip-chip on glass solutions for display applications.

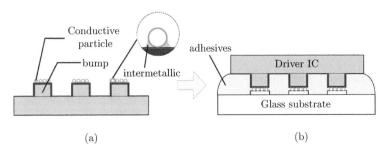

Figure 4.10   Schematics particle on bump

(a) particle on bump by intermetallic connection (b) chip on glass by insulated adhesive

Figure 4.10 shows schematically the main process of POB technology. First, one layer of conductive particles is placed on the bump surface of a driver IC with a proper particle density to fulfill the contact resistance requirement, and then, the particles are joined with the bump through intermetallic formation (Figure. 4.10a). Second, the driver IC is assembled on the glass substrate of a liquid crystal display LCD panel with an adhesive by thermal press (Figure. 4.10b).

The key process in POB technology is how to realize the intermetallic connection between the particles and a bump. Since most of the driver ICs in display modules are fabricated with the electroplating-Au bumps, and the conductive particles are almost Ni/Au coated polymer spheres, the Au-Sn metallurgy has been adopted to realize the intermetallic connection between particles and bumps. At first, a thin layer of tin was deposited on the top of the Au bumps. Conductive particles 4 μm in diameter with Ni/Au coating on polymer core were manually placed on the tin surface. Then the whole structure was heated in a thermal chamber to the Au-Sn eutectic temperature. In the heating process, the whole annealing time is about 10 minutes.

Figure 4.11 shows those conductive particles trapped under the bumps, which were captured from the backside of the glass substrate by optical microscope. Obviously, there is no agglomeration of conductive particles between the adjacent bumps in the POB image. This agglomeration is more and more critical to the electrical shorts for the conventional ACF-COG bonding as the bump pitch becomes finer and finer. From the image, we can find that the particles under the bump have a crushed shape similar to the normal ACF-COG particles, which indicates that the particles in both cases provide the same spring effect.[36]

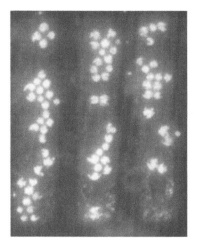

Figure 4.11   Conductive particles trapped under the bumps for POB-COG bonding

4.7.2   3D Interconnection

Demand for high-speed, high-density, small size, and multifunctional electronic devices has driven the development of 3D packaging, in which different functional devices are expected to integrate into one package.[37]

3D interconnection for SIP can be classified into five types, including wire bonding for stacked die, vertical interconnecting for package on package (POP), edge interconnection, 3D interconnection for embedded wafer level packaging, and through silicon via (TSV).[38,39] The first two approaches are already in mass production, while the rest remain for more investigation.

Wire bonding stacked die is a chip scale package (CSP) that has a small form factor. The traditional wire bonding method is used to connect pads on different dies. Spacers are added in between chips, if there is not enough height for wire bonding.[40] Wire bonding is a mature technology, with fast bonding speed and low cost. However, long loop wires have large resistance and other parasitic effects that prevent them from attaining high performance usage. In addition, heat dissipation is another problem for wire bonding. Thus, wire bonding technology is usually used to stack memory chips, including flash memory, DRAM (Dynamic Random Access Memory), and some other type memories. T. Watanabe presents a recent wire bonding stacked die result that DRAM chips are stacked together, with wires of a rather high density.[41]

In order to solve the reliability problem, a modified POP using through mold via (TMV) as the bottom module comes out.[42] TMV has a full molded bottom module with vias providing electrical interconnect. As a result, the bottom module has a closer warpage curve with the top module, which can reduce the total warpage. Figure 4.12 shows a sketch of POP with the TMV bottom module.

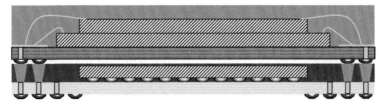

Figure 4.12   Sketch of POP with TMV bottom module

Edge interconnection is an emerging technology. Conductive adhesive is used to connect each die stacked together. A needle with conductive adhesive on its tip moves along a triangle path to extrude conductive adhesive between pads of different chips. The advantage of edge interconnection is that it can provide shorter interconnection and is supposed to be low cost. In addition, it can be stacked up as high as 128 dies.[38] However, edge interconnection can only be used in chips with the same size, such as memory chips. An example of edge interconnection was presented by Vertical Circuits.[43]

Through silicon via is believed to be the best 3D die stacking method.[44] TSV has the shortest interconnection distance, which means it has the fastest speed. Besides application in memory chip stacking, TSV can also be used for heterogeneous integration, in which logic, memory, MEMS, optical, and RF are stacked together. With embedded microfluidic channels, TSV can also solve heat dissipation problems. Figure 4.13 shows a sketch of dies stacked up with TSV.

TSV fabrication process includes "via first" and "via last" approaches. Via first means fabrication TSV before CMOS process, while the via last approach means fabricating TSV right after the CMOS process, and before the packaging process. The via first approach usually forms small vias, about 5 μm or less. Polysilicon or tungsten is used to fill these vias, providing electrical connection from the front side to the back side. The via last approach usually focuses on larger TSVs, from 10 μm to 100 μm in diameter. Electrical plating copper is widely used to fill these large TSVs, for its low resistance, high throughput, and low cost.

The CMOS image sensor (CIS) is the first commercial product using TSV technology. A CIS with TSVs can be directly mounted on the PCB of a consumer mobile product through the back side electric contact, which greatly reduces its size.[45] Figure 4.14 shows a sketch of the CIS module with TSV.

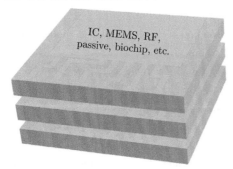

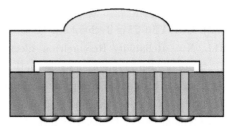

Figure 4.13   Sketch of TSV stacked dies   Figure 4.14   Sketch of CIS with TSV

Questions

(1) What are the characteristics of WB, FC, and TAB interconnections? In what areas are they mainly used?

(2) What are the characteristics of 3D interconnection? What do the main methods for 3D interconnection include?

(3) What parameters should be taken into account when choosing solder? Why should we develop lead-free solder?

References

[1] M.B. Tian. Electronic Package Engineering. Beijing: Tsinghua University Press, 2003.

[2] T. He. Actuality and Trend of Development of Wire Bonded Interface. Equipment for Electronic Products Manufacturing, 10(2004): 12–14, 77.

[3] Editing group of Production Broach, China Electronics Society. Microelectronic Packaging Technology. University of Science and Technology of China Press, 2003.

[4] Y.S. Li. Technology Analysis of Wire Bonder. Equipment for Electronic Products Manufacturing, 3(2004): 1–4.

[5] W.J. Chen, Xu Yang, Shuanke Yang, et al. Research Status of Package and Interconnect Structure Based on Integrated Power Modules. Application of Electronic Technique, 4(2004): 1–4.

[6] C.S. Wang and R.M Zhang. The influence of ultrasonic parameters in wire bonding performance. Electronic Component News, 2(2002): 27–31.

[7] F.M. Cheng. New Progress of Tape Automated Bonding. Hybrid Microelectronics Technology, 1(1990): 39–45.

[8] Y.S. Guo and H.K. Yu. TAB Carrier Technology. Application of IC, 4(2003): 17.

[9] S.J. Li and D.Y. Gao. The Introduction for TAB Assembly Process. Microprocessors, 1(1995): 43–45.

[10] Hoffman, P. TAB Implementation and Trends. Solid State Technology, 6(1988): 85–88.

[11] Y.X. Kuang and L. Liu. Technology of Bumped Tape Automated Bonding. Hybrid Microelectronics Technology, 2(1991): 11–15.

[12] M. Lou. Tape Automated Bonding and its Application. Computer Engineering and Applications, 11(1992): 52–57.

[13] B. Li, H. Wang, D. Wang, et al. Solder Bump Flip-Chip Bonding Technology. Semiconductor Information, 2(2000): 40–44.

[14] J.J. Lai, X.Q. Chen, H. Zhou, et al. Laser locally heating and bonding for microsystem packaging. Micronanoelectronic Technology, 7(2003): 257–260.

[15] W.H. Pei, H. Deng, and H.D. Chen. Flip chip bonding technology used in modern microphotoelectron package. Micronanoelectronic Technology, 7(2003): 231–234.

[16] Y.H. Cen. Flip Chip Technology. Hybrid Microelectronics Technology, 3(1998): 8–11.

[17] B.L. Xu. Reliability Research on Electronic Packaging. Doctoral Dissertation, Shanghai Institute of Micro-system and Information System. 2002.

[18] Q. Zhang. Research of underfill delamination of flip chip. Doctoral Dissertation. Shanghai Institute of Micro-system and Information System. 2001.

[19] Kari Kulojarvi and Jorma Kivilahti. A new under bump metallurgy for solder bump flip chip application. Microelectronics International, 15.2(1998): 16–19.

[20] M. Kleina, H. Oppermannb, R. Kalickib, et al. Single chip bumping and reliability for flip chip process. Microelectronics Reliability, 9(1999): 1389–1397.

[21] Kristiansen, H. and Liu, J. Overview of conductive adhesive interconnection technologies for LCDs. IEEE Transactions on Components, Packaging and Manufacture Technology, 21.2(1998): 208–214.

[22] Hop S., Jackson K. A., Lic Y., et al. Electronic Packaging Materials Science VI. Materials Research Society, 1992. 506 Keystone Drive, Warrendale, PA, USA

[23] Heinrich, W., Jentzsch, A., and Baumann, G. Millimeter-wave characteristics of flip-chip interconnects for multichip modules. IEEE Transactions on Microwave Theory and Techniques, 46.12(1998): 2264–2268.

[24] A.E. Ruehli and A.C. Cangellaris. Progress in the methodologies for the electrical modeling of interconnects and electronic packages. Proceedings of the IEEE, 5(2001): 740–771.

[25] Y.D. Hu and B.C. Yang. The Varieties of 3-D MCM. Electronic Components and Materials, 4(2002): 23–27.

[26] Tummala, R.R. Fundamentals of Microsystems Packaging. New York: McGraw-Hill, 2001.

[27] Harman, G.G. and Johnson, C.E. Wire bonding to advanced copper, low k integrated circuits, the metaldielectric stacks, and materials considerations. IEEE Transactions on Components and Packaging Technologies, 25.4(2002): 677–683.

[28] Vijay K. Varadan, K.J. Vinoy, and K.A. Jose. RF MEMS and Their Applications. West Sussex: John Wiley and Sons Ltd, 2003.

[29] C. Chiu, K.C. Chang, J. Wang, and C.H. Lee. Challenges of Thin Core Substrate Flip Chip package on Advanced Si Nodes. Proceeding In: of the 57th Electronics Components Technology Conference (ECTC), 2007.

[30] Ying-Hui Wang and Tadatomo Suga. 20 μm Pitch Au Micro-bump Interconnection at Room Temperature in Ambient Air. In: 2008 Electronic Components and Technology Conference, 2008: 944–949.

[31] Ankur O. Aggarwal, P. Markondeya Raj, Baik-Woo Lee, Myung Jin Yim, Mahadevan Iyer, C. P. Wong and Rao R. Tummala. Thermomechanical Reliability of Nickel Pillar Interconnections Replacing Flip-Chip Solder Without Underfill. IEEE Transactions on Electronics Packaging Manufacturing, Vol. 31, No. 4, October 2008: 341–354.

[32] P. Muthana, M. Swaminathan, R.R. Tummala, V. Sundaram, L. Wan, S. Bhattacharya, and P.M. Raj. Packaging of multicore processors: Trade offs and potential solutions. In: Proc. Electron. Compon Technol. Conf., 2005: 1895–1903.

[33] J.U. Knickerbocker, et al. Development of next-generation system-on-package (SOP) technology based on silicon carriers with fine-pitch chip interconnection. IBM J. Res. Dev., vol. 49, no. 4/5, 725–753, 2005.

[34] International Technology Roadmap for Semiconductors. 2003. www.itrs.net/Links/2003 ITRS/Home2003.htm

[35] Soga, I. Kondo, D., Yamaguchi, Y. et al. Carbon nanotube bumps for LSI interconnect, 2008 Electronic Components and Technology Conference.

[36] Lei Jia, Zhiping Wang, and Zhenhua Xiong. Particle on Bump (POB) technique for ultra-fine pitch chip on glass (COG) applications. In: Proc. 8th International Conference on Electronic Packaging Technology 2007: 1–4.

[37] T. Jiang and S. Luo. 3D Integration-Present and Future. Electronics Packaging Technology

Conference, 2008, 373–378.

[38] A. Yoshida, et al. A study on package stacking process for package-on-package (PoP). Electronic Components and Technology Conference, 2006: 825–830.

[39] D. Shi. Comparisons of Various 3D Packaging Technologies. In: Proc. Advanced Packaging Technologies Consortium Workshop, 2009.

[40] M. Karnezos. 3D packaging: where all technologies come together. In: Electronics Manufacturing Technology Symposium, 2004, 64–67.

[41] T. Watanabe. The Memory Packaging Strategy with Sophisticated 3D Technology. International Conference on Electronics Packaging, 2009.

[42] J. Kim, et al. Application of through mold via (TMV) as PoP base package. Electronic Components and Technology Conference, 2008, 1089–1092.

[43] L.D. Andrews, T.C. Caskey, and S.J.S. McElrea. 3D Electrical Interconnection Using Extrusion Dispensed Conductive Adhesives. In: Electronic Manufacturing Technology Symposium, 2007, 76–80.

[44] J.U. Knickerbocker, et al. 3D Silicon Integration. In: Electronic Components and Technology Conference, 2008, 538–543.

[45] D. Henry, et al. Through Silicon Vias Technology for CMOS Image Sensors Packaging: Presentation of Technology and Electrical Results. In: Electronics Packaging Technology Conference, 2008, 35–44.

CHAPTER 5

Device-level Packaging

5.1 Introduction

5.1.1 Basic Concepts

Device-level package (DLP), also called single chip package, encapsulates a single circuit or chip to provide it with the necessary electronic connections, mechanical support, thermal management, protection from harmful elements, and an interface for future applications. Packaging for two or more chips in a single module is usually called multichip package or multichip module.

A typical process of device level packaging is shown in Figure 5.1, including dicing, mounting, bonding, encapsulating, marking, leads or solder ball fabrication, and unification of the finished product.

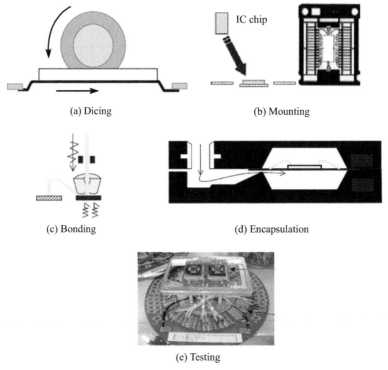

(a) Dicing

(b) Mounting

(c) Bonding

(d) Encapsulation

(e) Testing

Figure 5.1   Typical process of device-level packaging

5.1.2 Status and Role

As a very important part of microsystem package technology, DLP guarantees a normal chip proper operation and communication between the chip and the environment. There are many kinds of DLPs, all of which should have the following basic features:

(1) Provide a reliable electronic signal I/O transmission to provide power, grounding, signals, and other stable and reliable power supply.

(2) Satisfy different requirements of system package for the device to ensure effective signal transmission and power supply after second-level packaging.

(3) Assemble the device on substrate during higher level packaging through insertion mounting, surface mounting, or other interconnection assembly approaches.

(4) Provide effective thermal dissipation to send out the heat produced by the packaged devices.

(5) Provide effective mechanical support and isolation protection to avoid the damage from environmental vibration, mechanical force applied during packaging, moisture, and caustic gases to devices.

(6) Provide flexible physical transition to fit fine chip to substrates of different sizes.

(7) Provide a low-cost packaging scheme while satisfying system requirements and realizing the designed performance.

Therefore, when choosing package materials, low cost, good electrical performance, such as low dielectric constant, excellent thermal conductivity, and sealing performance should be considered.

5.1.3   History

The history of electronic device packaging is briefly shown in Figure 5.2. For most chips, the expected functions can only be achieved after they are packaged. Packaging is also called encapsulation when using an organic material, usually low-temperature polymers; encapsulation is nonhermetic sealing. Inorganic materials, such as metals, ceramics, and glasses, can also be used for packaging, in which case it is called sealing. Sealing is hermetic, while it can be vacuum, or filled with nitrogen gas or inert gases inside the sealed chamber. Sealing is reliable but expensive. With the development of packaging techniques and materials, the reliability of encapsulation is greatly improved, which makes it widely used in many applications. High-cost sealing is replaced by low-cost quasihermetic encapsulation in those areas.

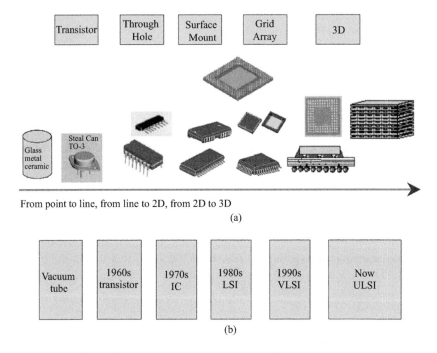

Figure 5.2   Development of electronic device and package

Encapsulation now accounts for more than 95% of all device packages. But in some specific areas, such as military, space, and underwater applications, sealing is still necessary.

In 1947, the first semiconductor transistor was invented. That's the beginning of electronic packaging history.

In the 1950s, TO-type shell packages were developed with three electric pins and metal-glass sealing. During the same period, ceramic casting was invented, which provided a process lead for multilayer ceramics. In 1958, the first integrated circuit came out. Though metal glass packaging still took the leading role, the development of the multiwire package began to attract much more attention. With the development of the integrated circuit from small-scale to middle-scale then to large-scale, the scale of integration was getting higher and higher and more and more I/O pins were required, thus promoting the development of multilayer ceramics.

In the 1960s, the double in-line package (DIP) with a ceramic shell was born, that is, ceramic double in-line packaging (CDIP). With excellent electrical and thermal performance and high reliability, CDIP was highly favored by integrated circuit manufacturers and developed quickly. It soon became a leading product in the 1970s. Later, plastic DIP was developed. It costs little to mass produce. Now it is still used widely in the low-end product market, with the pin number less than 84 and pitch usually 2.54 mm.

In the 1970s, with the development of SMT, a series of electronic package products for the SMT process were developed, such as leadless ceramic carrier, plastic leaded chip carrier, and quad flat package (QFP), and were put into commercial production in the 1980s. For its high density, small pitch, low cost, and easy surface mounting, plastic QFP became a leading product in the 1980s.

In the 1990s, more and more I/O pins were required for packaging ultra-large-scale integrated circuit and pitch was getting smaller and smaller. Since then, electronic packages developed include quad flat package, such as QFP with 356 pins and 0.4 mm pitch, area array package, such as pin grid array (PGA) with 750 pins and 1.27 mm pitch, and ball grid array (BGA), over 3000 pins and 0.8 mm pitch. Now the development of advanced packaging for high density I/O is still booming.

Since the beginning of the 21st century, different CSPs of smaller size packages have become the emphasis of research and development. Packaging, at the same time, is moving in the direction of high density and high performance, such as 3D stack package, SIP, or SOP.

Various packaging methods are designed for different chips, which should satisfy special requirements of assembling and next level package or mounting. Depending on the package materials used, packaging can be divided into metal packaging, plastic packages, and ceramic packages, including metal ceramic package. These three packages will be discussed below.

5.2   Metal Packaging

5.2.1   Concept of Metal Packaging

Metal packaging uses metal as a shell or base to seal and protect a chip or chips. The I/O leads for transmitting electric signal, power, and ground in metal package is mostly sealed with glass metal or metal ceramic material.

Because of its good performance in thermal dissipation and electromagnetic shielding, metal packaging is usually used in highly reliable or customized hermetic packages. The main application of packaging modules, circuits, and devices includes packaging microwave multichip modules and hybrid circuits, power devices, ASIC, photo-electronic devices, and other specific devices.

5.2.2   Advantages of Metal Packaging

Metal packages feature high precision, strictly controlled size, low cost, batch production capability, excellent performance, wide applicability, high reliability, and large packaged free space.

With various methods and flexible processing technologies, metal packages have good compatibility with some components, such as hybrid integrated A/D or D/A modules, applicable for packaging of low I/O single chips or multichips. In addition, it is applicable for packaging of MEMS, RF, microwave, optoelectronic, surface acoustic wave, and large power devices. It can meet the requirements of packages in small batch production with high reliability. Moreover, in order to provide good thermal dissipation, the metal structure also serves as a heat sink in most metal packages.

5.2.3   Process of Metal Packaging

A typical process of metal packaging is shown in Figure 5.3. The first step starts by making the metal lid and shell of metal package. Then electrodes on the shell are made to provide input and output ports for power supply and electrical signal transmission. The electrodes are sealed and insulated with the metal/glass/metal sealing method. After the thinning and dicing processes, the chip will be mounted on the shell with the attaching and wire bonding steps. The last step is capping.

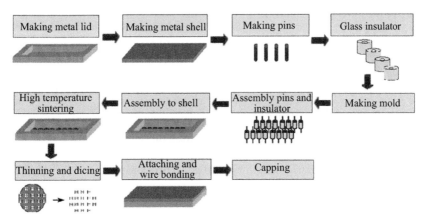

Figure 5.3   A typical process of metal packaging

Metal parts should be baked before the chip assembly to remove the gases or moisture out of the surface of the metal to reduce device defects caused by corrosion. During assemblly, high temperatures cannot be kept for a long period of time. Instead, a temperature curve should be followed to reduce the thermal influence of the postprocesses on the preprocess etch steps.

Capping is a special technique used in metal packages. The usual capping techniques include resistance welding seam welding, pulse heat fusing, laser welding, and solder sealing or any other microjoining method. During the capping process, there should be no gap between the sealing surface of the package cover and shell, and the sealing parts should be precisely aligned, since sealing defects of the devices Otherwise there may be. Capping is usually done in a dry protective environment, such as nitrogen, to reduce moisture and other harmful gases.

Seam welding is one of the highly reliable capping technologies.[1] Materials like cover plates or lids may greatly affect the hermetic performance and the leaking rate of hermetic end products. High-quality cover plates for seam welding should have the following features:

(1) The thermal expansion coefficient should be similar to that of shell and solder rings and close to that of the ceramic substrate. (2) The melting point of bonding parts should be as low as possible. (3) They should have excellent corrosion resistance. (4) They should have a small tolerance in size. (5) They should have a uniform and smooth surface without any burr or contamination.

The base materials mostly used now are alumina ceramics and Kovar alloy. The metal solder ring that can match ceramic's coefficient of thermal expansion (CTE) is Kovar alloy or 4J42 ferro-nickel alloy.

Kovar alloy's melting point is 1460°C. In order to reduce the melting point of the lid, nickel-phosphorus alloy is plated onto the bonding surface, and a low bonding temperature of 880°C is thus achieved.

Tin-lead solders are widely used for bonding; alloying additions such as indium and silver are sometimes added to improve the strength or fatigue resistance. Use of bismuth-tin alloys for sealing with a lower melting point than eutectic tin-lead solder has also been suggested. And in addition, Au-Sn is a common bonding material, especially for solder bonding between two materials with similar CTEs. Eutectic (80:20) Au-Sn alloy soldering is also called brazing. In furnace sealing, the typical reflow time is 2–4 minutes above the eutectic temperature of 280°C, with a peak temperature of about 350°C. When bonding two parts with mismatched CTEs, fatigue failure may appear after thermal cycling tests. In addition, Au-Sn is also a fragile material and can only stand low stress.

5.2.4   Traditional Metal Packaging Materials

To provide chips with mechanical support, electric interconnection, thermal dissipation, and environmental protection, metal package materials should meet the following requirements:

(1) Low CTEs similar to that of chips or ceramic substrates to reduce or avoid thermal stress.

(2) High thermal conductivity to provide thermal dissipation.

(3) Good electrical conductivity to reduce transmission delay.

(4) Good EMI/RFI shielding performance.

(5) Low density, high strength and rigidity, and good processing or molding characteristics.

(6) Good characteristics for plating, welding and corrosion resistance, easy to attach chip onto substrates reliably and seal lid with shell hermetically.

(7) Low cost.

Selection of metal materials may directly affect the quality and reliability of metal packages. The common-used materials are Al, Cu, Mo, W, steel, Kovar alloy, CuW (10/90 or 15/85), Silvar™(Ni-Fe alloy), and CuMo (15/85). All of these materials have good thermal conductivity with higher CTEs than that of silicon. The density, CTE, and thermal conductivity of some materials are listed in Table 5.1.

Table 5.1   Main performance of commony used package materials

Materials	Density (g·cm^{-3})	CTE ($\times 10^{-6}$K^{-1})	Thermal conductivity [W/(m·K)]
Si	2.3	4.1	150
GaAs	5.33	6.5	44
Al$_2$O$_3$	3.61	6.9	25
BeO	2.9	7.2	260
AlN	3.3	4.5	180
Cu	8.9	17.6	400
Al	2.7	23.6	230
Steel	7.9	12.6	65.2
Stainless steel	7.9	17.3	32.9
Kovar alloy	8.2	5.8	17.0
W	19.3	4.45	168
Mo	10.2	5.35	138

5.2.5   Novel Metal Packaging Materials

Traditional metal packaging materials are single metals or alloys except Cu/W and Cu/Mo. These materials all have their own limitations and are not able to meet the requirements of the modern package technology development. In recent years, a lot of metal matrix composites (MMCs) have been developed. They are composites with Mg, Al, Cu, and Ti or metal alloys such as the matrix, and they have granules, whiskers, short fiber, or continuous fiber as reinforcement. Compared with traditional materials for metal packaging, MMCs have the following advantages:

(1) It is possible to change the thermal physical properties of materials by adjusting varieties of reinforcement parts, volume fraction, fiber orientations, or matrix alloy to meet the requirements of package thermal dissipation and even simplify the package design.

(2) Flexible material manufacturing, especially direct molding, can prevent expensive processing costs and material waste caused in processing.

(3) The specially developed low-density and high-performance MMCs are quite likely to be used for aviation and avionics applications.

The main composite materials used for microsystem packages are composites of Cu-matrix and Al-matrix thermally matching with Si material. Their properties are shown in Table 5.2.

Table 5.2   Properties of Cu-matrix and Al-matrix composites

Metal-matrix	Reinforcement	Thermal conductivity $[\mathrm{W(m^{-1}K^{-1})}]$		CTE ($\times 10^{-6}\mathrm{K^{-1}}$)	Density (g·cm^{-3})
		$X - Y$	Z		
Cu	$\pm 2°$ SRG	$840(x), 96(y)$	49	$-1.1(x), 15.5(y)$	3.1
Cu	$\pm 11°$ SRG	$703(x), 91(y)$	70	$-1.3(x), 15.5(y)$	3.1
Cu	$\pm 45°$ SRG	$420(x), 373(y)$	87	$1.2(x), 3.6(y)$	3.1
Cu	$0°, 90°, 0°$	$415(x), 404(y)$	37	$5.3(x), 5.4(y)$	3.1
Al	2D Fabric 1	280	NA	2.8	2.3
Al	3D Fiber Mat 1	187	74	10.4	2.5
Al	3D Fiber Mat 2	226	178	5.5	2.3
Al	MMCC 3D-2	222	100	5.0	2.3
Al	MMCC 3D-1	189	136	6.0	3.1

Along with the development of electronics packages for high performance, low cost, low density, and system integration, requirements of metal packaging materials are getting higher and higher. MMC will play a more and more important role. Therefore, the research and application of MMCs will become a hot topic in the future.

5.2.6   A Case Study of Metal Packaging[2]

Most MEMS devices need vacuum packaging to ensure a good working environment for their movable parts. Metal packages with excellent sealing features are the first choice for high-performance MEMS vacuum packages.

Figure 5.4 is the schematic diagram of the MEMS vacuum package. The cap and house of the package are made of Kovar material. Then electrodes are formed by sintering the glass tubing with Kovar wire throughout the house. A house is fabricated, where low-temperature bonding material is used to form a hermetic seal between the cap and house.

The process flow of low-temperature sealing is as follows: first fix MEMS chip and low-temperature getter inside the house and cap, respectively, and introduce low-temperature solder inside the slot of the house. Then electrically interconnect the chip with feedthrough electrodes of the house by the wire bonding process. Second, assemble the house and cap into their positioning frames and move them into the vacuum chamber. Third, after pumping the vacuum chamber to a set pressure, heat up the cap at 400–500°C for several minutes based on the activation specification of a low-temperature getter to activate the getter and melt the bonding solder. Fourth, reduce the temperature and keep it around 170°C for several hours to thoroughly degas the packaged parts. Finally, seal the cap with the house by sinking the port of the house into the liquidized solder in the slot of the cap, and reduce the temperature gradually in a vacuum environment.

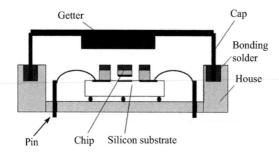

Figure 5.4   Schematic diagram of MEMS vacuum packaging

5.3   Plastic Packaging

5.3.1   Concepts and Features of Plastic Packaging

A plastic package uses resin material to encapsulate a semiconductor device or circuit chip and is usually considered nonhermetic package. It features simple process, low cost, and ease of mass production. Plastic packaged products account for almost 95% of the IC package market, with reliability continually improving. They are widely used in applications with frequencies under 3 GHz.

Standard plastic packaging materials mainly include 70% fillings such as silicon dioxide, 18% epoxy resin, added curing agent, coupling agent, mold release agent, flame retardant agent ,and coloring agent. The amounts of ingredients depend on parameter requirements in product application, such as CTE, dielectric constant, sealability, hygroscopicity, strength, etc. and factors like improving strength, lowering prices, etc.

5.3.2   Working Process and Procedures of Plastic Packaging

Plastic packaging, if not specifically indicated, in general, refers to transfer molding package. As shown in Figure 5.5, the main process steps include wafer thinning, slicing, chip mounting, wire bonding, transfer molding, post solidifying, deburring, solder plating, trimming, and lead forming, as well as marking.

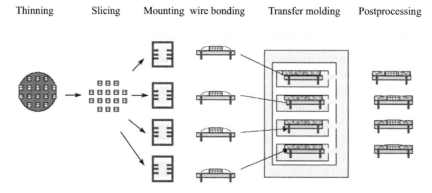

Figure 5.5   Typical process of plastic packaging

Transfer molding is a technique for forming thermosetting plastics. In this process, raw material is first heated in a heating chamber, then it is pushed into a preheated cavity for curing.

The low cost of this technique makes it suitable for large-scale production. Therefore, it is currently the most commonly used packaging method in the semiconductor industry.

In transfer molding, plastics should have a high fluidity below the curing temperature and a high curing rate above the curing temperature. Phenolics, melamine-formaldehydes, and epoxy resins meet these requirements.

Transfer mold packaging has the following advantages:

(1) Products have fewer wasted parts, and thus the postprocessing effort is reduced.

(2) Products with fine or fragile inserts or holes can be produced, with the inserts or holes being kept in the right locations.

(3) Products have consistent properties, accurate size, and high quality.

(4) The molds have less wear and tear, in comparison with injection molding.

Disadvantages:

(1) Mold cost is high.

(2) Plastics waste is high.

(3) Anisotropical behavior can resalt from the fiber orientation in fiber reinforced plastic.

(4) Sometimes, poor adhesion of the plastic will reduce the strength of products.

The process before encapsulation with plastic material is called the assembly step or preprocess, while the process after encapsulation is called postprocess. Preparations before plastic packaging include manufacturing of the chip to be packaged, preparing for the mold, and fabrication of leadframe.

5.3.3   Types of Plastic Packaging

From the application point of view, plastic packages can be sorted into three groups, including through-hole, surface-mount, and TAB. Common plastic packages for ICs include several types, as shown in Figure 5.6.

PDIP: plastic double in-line package

PLCC: plastic leadless chip carrier

PSOP: plastic small-outline package

PQFP: plastic quad flat package

(a) PDIP   (b) PLCC   (c) PSOP   (d) PQFP

Figure 5.6   Typical plastic package devices

PQFP is more applicable for packaging chips with large number of I/O pins. The mold temperature during the package is about 150°C with pins usually pressed into the shape of wings. This package is applicable for surface mounting on PCB.

PPGA: plastic pin grid array

A Ni-plated copper radiator can be mounted on the top of a PPGA to improve its thermal conductivity.

PBGA: plastic ball grid array

In 1991, PBGA based on organic resin substrates was available. It was first used in computers, radio receivers, ROM, and SRAM.

TBGA: tape ball grid array

5.4 Ceramic Packaging

5.4.1 Profile of Ceramic Packaging

Like the metal package, the ceramic package is also a hermetically sealed package with low cost. Materials mainly used for this type of package are Al_2O_3 with a CTE of $6.7 \times 10^{-6}°C$.

With the development of ceramic tape-casting, the ceramic package becomes more flexible in shape and function. IBM ceramic substrate technology, for example, can manufacture a patterned substrate with over 100 layers, and it is possible to integrate passive components, such as resistors, capacitors, and inductors, in ceramic substrates to realize high-density packages. In 2004, Huawei Co. in China developed a ASIC with 0.13 μm in feature size and 40 million gates, in which ceramic BGA with 1120 pins was introduced.

The main features of ceramic packages are

(1) They have good hermetization and high reliability.

(2) They have excellent electrical performance, are able to realize multisignal, ground and power I/O structure, and are capable of integrated packages for complicated devices.

(3) They have less size limitation and cost reduction with heat management, due to its good thermal conductivity.

(4) They have low size precision in the sintering process and a high dielectric coefficient, as well as high cost compared with plastic packages.

Ceramic package's coverage in the high-end package market is increasing year by year because of its excellent performance and varied applications in aviation and avionics, military, and large-scale computer fields.

5.4.2 Process of Ceramic Packaging

The process of ceramic packaging can be divided into two steps. First, prepare the package base as shown in Figure 5.7, which includes molding the raw ceramic substrate, metalizing and electroplating to form electrodes, layer laminating, and sintering.

Figure 5.7 Samples of a ceramic package base

Like the procedure shown in Figure 5.1, the second step to complete a package includes chip attaching, bonding, strengthening (if necessary), and capping. Capping of ceramic hermetic packaging can be welded, soldered, brazed, or sealed by glass sealing. To facilitate the sealing of the ceramic substrates by soldering or welding, a metal seal ring should be provided on the substrate surface, which is formed by thick film, cosintered copper, or tungsten metallurgy. The seal frame is then electroplated, and a metal lid is attached by soldering or welding. The large throughputs, high yields, and reliability associated with the welding technique are spurring a change from glass sealing to welding for ceramic packages. The major considerations in selecting a sealing method are the availability of equipment and the cost of the hybrid circuit. Welding is more economical because of its high productivity and high reproducibility. Solder or braze sealing is commonly employed when it is required to dissemble and reseal the lid. Of these, the most popular method of hermetic sealing is welding.

5.4.3   Types of Ceramic Packaging

Ceramic packages mainly include metal ceramic packages and common ceramic packages with great varieties. The former are mainly applied in various packages for coaxial or carrier types of single devices and micro/mm wave integrated circuits, while the latter are widely used in various integrated circuit packages.

CDIP: ceramic double in-line package

LCC: leadless chip carrier

CQFP: ceramic quad flat packaging

Ceramic array packages developed rapidly, and a variety of package solutions emerged:

CPGA: ceramic pin grid array

FC-CBGA: flip-chip ceramic ball grid array

FC-CCGA: flip-chip ceramic column grid array

C-CSP: ceramic chip scale package

5.4.4   An Example of Ceramic Packaging Application—High Brightness LED Package[3]

By making use of the ceramic package's advantages of good thermal conductivity and low cost, Kyocera developed ceramic package solutions for white light emitting diode (LED) and blue LED high brightness and started batch production in early 2003. Owing to the use of alumina and aluminum nitride with better thermal properties in the package materials, compared to the existing LED resin material, the thermal properties and heat endurance are improved. This type of ceramic package is mainly used to backlight mobile phones, and white LED is used in lighting equipment. Although the existing package is fully suitable for packaging the white LED used in the phone backlight, the LED chip size became greater than the backlight and needs a bigger current to drive, the result of which is that a high heat dissipation package, such as a ceramic package is necessary for the new structures. This is because the resin material dissipates heat poorly; therefore, if working under high temperatures, LED chips and the resin material will age rapidly.

5.5   Typical Examples of Device-level Packaging

According to the materials used for DLP, metal, plastic, and ceramic packages were introduced as above. From the technical perspective, device-level packaging has developed from DIP, QFP, PGA, BGA, to CSP, with increasing package efficiency, higher adaptibility, better heat endurance, better reliability, and more convenience for end users. Owing to limits to the scope of the book, it is impossible to give a detailed introduction of them one by one; only the most commonly used—DIP, BGA, and CSP—are introduced here.

5.5.1   DIP

DIP, that is, dual in-line package, is usually used in DRAM Dynamic Random Access Memory and most middle- and small-scale integrated circuits. The pin number of a DIP is usually no more than 100. Figure 5.8 shows one example of its application.

DIP features easy operation on a PCB with through-hole assembly and a high volume ratio of package device to chip.

DIP has two rows of pins that are to be inserted in the socket of the DIP structure on an assembling board with the same hole numbers and in geometric arrangement for soldering. DIP was once quite popular during the time when memory devices were directly inserted in the motherboard of computers. Special care must be taken when a DIP is inserted or taken away from the socket.

Figure 5.8   Photo of a DIP packaged MEMS sensor

There are many kinds of DIP structures, such as multilayer ceramic DIP, single-layer ceramic DIP, and lead frame DIP, including glass ceramic package, plastic encapsulation, and ceramic low-melting glass package. A derivative form, shrink DIP (SDIP), has a pin density six times more than that of a normal DIP structure.

Some factors to be considered in the future of DIP packaging follow: the first is to increase the size efficiency of the chip by improving the ratio of chip size to package size to be as close to 1:1 as possible; the second is to maximize pin length to reduce the delay and guarantee the distance between the pins to avoid interoperability and improve performance; the third is to reduce the package thickness to improve thermal dissipation.

As discussed above, the chip: package area ratio is an important indication of the advancement of packaging technology. The closer the value to 1, the more advanced the technology is. Take a CPU using 40-I/O-pin PDIP as an example, its chip/package area ratio $= (3 \times 3)/(15.24 \times 50) = 1 : 86$, very far from the expected value of 1. The package size is much bigger than the chip size, and so the package efficiency is low. That means the package covers too much effective mounting area. In the 1980s, chip carrier packages appeared, including leadless ceramic chip carrier (LCCC), plastic leaded chip carrier (PLCC), small outline package (SOP), and plastic quad flat package (PQFP). Compared with DIP, carrier packages greatly reduced package size and improved reliability as well. That's why Intel's CPUs for example, 80386, adopted PQFP during this period.

In the 1990s, with the development of integration techniques, improvements in equipment capability, and application of deep submicron processes, LSI, VLSI, and ULSI appeared one after another with improved integration scale of silicon chips, and brought higher requirements for packaging these IC chips because package technology may affect product performance. When the working frequency of an IC chip is higher than 100 MHz, traditional package solutions may cause "cross talk." In addition, traditional package technology becomes quite difficult to deal with when IC chips have over 208 pins. With the I/O pin number increasing rapidly, the power consumption increases as well. A new BGA product appeared to meet the demand of development. In addition to the use of QFP packages, BGA package technology is used in most of today's chips with large numbers of pins (such as the graphics chips, chipsets, etc.). Since the time of the emergence of BGA, it has been the best package choice for ICs such as motherboards, South/North Bridge chips etc., with high-density, high-performance, multipin packages.

5.5.2   BGA Packaging

Ball grid array (BGA) is a kind of high-density surface mount package.[4] At the bottom of the package, pins are all ball-like and arranged in rows of a grid similar to a lattice pattern; hence the name BGA. Figure 5.9 shows a typical BGA package and pads on a board.

Figure 5.9   Typical BGA package

The BGA package is a kind of ball array package with round or column solder joints formed by its I/O pins distributed in an array below the package. The large pitch and short lead length may avoid the coplanarity and warp caused by wires in fine-pitch devices. The structure of a typical BGA package is shown in Figure 5.10.

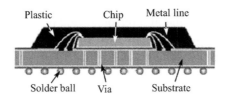

Figure 5.10   Typical structure of BGA package

BGA is one of the most popular ways to interconnect between IC and PCB. The most remarkable feature of BGA is that it is still applicable in the current SMT process for BGA devices with I/O numbers over 200. SMT's basic process is reflowing, which has proved suitable for assembling BGA devices. Although the time and temperature curve of BGA assembly corresponds with the standard curve of the SMT process, the special characteristics of this type of package should be kept in mind for this application. This is particularly important since solder joints of BGA, as opposed to the traditional devices, are usually below the device, between the device and the PCB. Therefore, the influence on the internal structure of materials is more significant than most traditional package forms. For this reason, it is necessary to decide reflow parameters with the measured value of BGA solder joint temperature as the reference value.

Features of the BGA package:

(1) Although the number of I/O pins increases, the pitch is wider than QFP, thus improving the assembly yield.

(2) The thickness is half that of QFP, while the weight is cut by more than three-fourths.

(3) Decreasing the parasitic parameters, such as signal transmission delay, greatly improves the frequency.

(4) Coplanar assembly soldering with high reliability is available.

(5) Like QFP and PGA, the occupied area of BGA is still oversized.

(6) Higher I/O numbers with larger pitch eliminates the problems of production cost and reliability for QFPs with high I/O numbers.

The BGA package family has a lot of members. They are not only different in size and I/O number, but also different in physical structure and package material.

1. Plastic BGA (PBGA)

PBGA is the most popular package in production. Figure 5.10 shows a typical plastic BGA package device. Its main features are

(1) Glass fiber and bis-maleimide-triazine (BT) resin substrate, with thickness about 0.4 mm.

(2) The chip is directly bonded onto the substrate.

(3) The chip is connected to the substrate by wire bonding.

(4) Molded plastic could encapsulate the chip, the interconnection wires, and the majority of the substrate surface.

(5) Solder balls (usually eutectic material) are soldered on the pads at the bottom of the substrate.

However, the area coverage of the substrate by plastic in this package should be considered carefully. In some cases of PBGA packaging, molding plastic will cover almost the whole substrate, while for others, the coverage is limited to a central area. This will affect the heat exposure of solder joints.

2. Ceramic BGA (CBGA)

The multilayer substrate with a metal interconnecting pattern is the most basic, material for CBGA. Its substrate is made of ceramic, and the package cover is made of aluminum. The sealing quality of this kind of package is the most influential factor to the thermal conductivity that can pass through the package. A variety of materials are used for the package "cover." An unfilled space may block the heat exposure of under-package solder joints. Though the power consumption is increased in this way, BGA can use controlled collapsed chip connection (C4) to improve its electrical thermal performance.

3. Enhanced BGA

The word "enhance" in enhanced BGA means enhancement of performance by adding some material to the structure. In general, the material added is metal that can improve IC thermal dissipation during operation. This is quite important because one of BGA's advantages is that it can provide a large number of I/Os for IC. Heat dissipation must be designed in detail for this kind of package, because chips packaged in this way usually generate much heat in a small space.

Special enhanced BGA, called super BGA (SBGA), features a structure with an inverted copper cavity attached to the top of the package to improve the thermal dissipation to the surroundings, as shown in Figure 5.11. On the bottom of the copper piece is a soft and thin substrate, which is used as the pads attached with solder balls along the side. The inner wires connect the substrate with the chip pads. The chip is plastic packaged from the bottom. Obviously, this structure will be greatly helpful for the thermal dissipation of the chip.

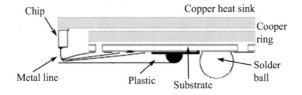

Figure 5.11   Cross section of SBGA

4. Tiny BGA Package

Tiny ball grid array (BGA) provides a ratio of chip: package area no less than 1:1.14. The application of this technology can increase the memory capacity by 2 to 3 times without changing the size of all DRAM memory in computers. Tiny BGA has replaced traditional thin small outline packaging (TSOP, which is the representative of second-generation memory package technology that appeared in 1980s and features making pins around the packaged chip) with smaller volume and better thermal dissipation and electrical performance.

Like the technology for packaging the microprocessor, TSOP was no longer applicable for the next generation memory with high frequency and high speed, along with the development of packaging technology. New packaging technology, with tiny BGA and bottom leaded plastic (BLP) as representatives, gradually developed. Memory products with tiny BGA packages reduce the volume to a third that of TSOP packages while keeping the same capacity. In tiny BGA packaged memory products, I/O leads are drawn from the center of the chip and can effectively reduce the distance of signal transmission and signal attenuation. Meanwhile, this type of package significantly improves not only the performance of anti-interference and anti-noise, but also the electrical properties of chip; tiny BGA packaging is very thin in size, with a height less than 0.8 mm, and the effective route from metal substrate to thermal sink is only 0.36 mm. Therefore tiny BGA memory has better thermal conductivity and is applicable for long-term operation systems with excellent stability. In the bottom leaded plastic (BLP) package, its ratio of chip:package area is 1:1.1. Not only are the height and area reduced, the electrical characteristics are also improved. In addition, production costs are also affordable. Hence BLP technology is becoming widely used.

5. Future Trends

Although BGA packaging is more advanced than QFP, its ratio of chip area to package area is still not ideal. Tessera had made some improvement based on BGA and developed another package technology named μBGA, whose chip: package area ratio is close to 1:1 with 0.5 mm pitch, really a big step from basic type of BGA. The typical structure of Tessera's μBGA is shown in Figure 5.12.

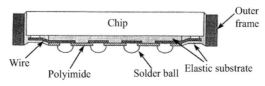

Figure 5.12   Tessera's μBGA

In September 1994, Mitsubishi Electric of Japan developed a new package structure with a chip: package area ratio to 1:1.1. Its package size is only slightly bigger than that of a bare chip. That means a package size is similar to the size of a single IC chip. This new package is called a chip scale package (CSP).

5.5.3   CSP

CSP, the chip scale package,[5] which was developed on the basis of TSOP and BGA, is a thin chip package.

CSP can reach a chip: package area ratio over 1:1.14, that is, about a third of the common BGA chip:package area ratio, and 1:6 of TSOP memory chip area. With a small volume, CSP is also thin in size. Its most effective route for thermal dissipation from metal substrate to thermal sink is only 0.2 mm, which greatly improves the reliability of the memory chip after long-term operation, reduces line impedance, and increases chip speed. Tassera's μBGA shown in Figure 5.12 is a typical CSP package structure. The outside frame is optional. Without the frame, its plane size is almost the same as the chip size.

CSP has better electrical performance and reliability than BGA and TSOP do. The number of pins in CSP is obviously greater than TSOP and BGA under the same chip area, which results in the great increase of I/O number. In addition, the CSP packaged memory chip effectively reduces signal transmission distance and signal attenuation and improves the disturbance and noisy resistance of the chip. CSP's signal transmission time is reduced by 15%–20% compared with BGA.

CSP is electrically connected with the outside through a solder ball array soldered to a substrate. Since the contact area between the solder ball and the substrate is larger, heat generated by the chip during the operation can be easily transferred to the substrate and then dissipated. CSP also can dissipate heat from the backside and has good thermal efficiency. Typical CSP thermal resistance is 35°C/W, while TSOP thermal resistance is 40°C/W. Testing results show that memory with CSP can conduct 88.4% of the heat to substrate, while TSOP memory can only conduct 71.3%. CSP also has a compact chip structure and a low circuit redundancy, so it can reduce a lot of unnecessary power consumption and further reduce chip power consumption and the operation temperature.

There are many types of CSP packages, which generally can be classified into the following four categorics, although they are different in design, materials, and applications:

1. CSP Based on a Customized Lead Frame

CSP based on a customized lead frame is a category that encompasses compasses many types of products, such as bump chip carrier (BCC), quad flat nonlead (QFN), lead-on-chip CSP (LOC-CSP), microstud array (MSA), and bottom leaded plastic (BLP) packages. LOC-CSP and BLP are introduced here. LOC-CSP is mainly used for chip expansions and system packages to keep the coverage of packages on the PCB unchanged.

LOC structure is introduced here. The structure permits increasing chip size without increasing package size and coverage of PCB. This package was developed in the early 1990s and began to be widely used in packaging memory devices, especially to meet the demands of large capacity DRAM. Such novel CSPs have a smaller mold plastic area on the top of the package, a wider shoulder for wire interconnection, and a very thin package size to make stacking packages more convenient.

Apart from the Si chip, LOC-CSP is composed of bonding Au wire, adhesive tape for attaching the chip, custom designed lead frame, and encapsulated mold plastic, as shown in Figure 5.13. To reduce package thickness further, liquid adhesive coating on the bottom of the inner wire can also be applied to replace adhesive tape.

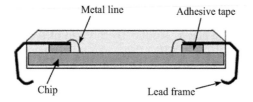

Figure 5.13   LOC-CSP structure

With short leads, the inductance and capacitance of leads of LOC-CSP can be neglected. Instead, the electrostatic capacitance should be considered because of the close distance between the lead frame and the wafer. Experiments found that the maximum variety of static capacitance caused by changing thickness of adhesives reaches 1 pF, which is too small to consider the electrical influence caused by the thickness of adhesives on LOC-CSP. The LOC-CSP also has good thermal dissipation, since it is thin and packaged without or with very thin plastic film on the bottom of the silicon chip. A thermal shock test also shows the package to be reliable.

BLP is one kind of leadless plastic encapsulation package, as shown in Figure 5.14. The chip is fixed onto the customized lead frame by adhesive. First-level and second-level interconnection use the wire bonding and electroplating flat pad, respectively. Owing to the short interconnection distance and compact structure, excellent electrical and thermal performance can be achieved. This kind of package makes application in high-speed devices with clock frequency over 1 GHz possible.

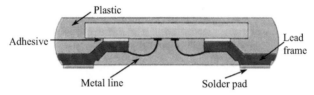

Figure 5.14   BLP package

The BLP package is designed for low I/O number applications, featuring small substrate coverage and thin shape. It is mainly used for packaging memory modules.

2. Flexible Substrate CSP

Flexible substrate CSPs use the flexibility of a substrate to redistribute the electrode pads of a chip featuring finer pitch onto the wider pitch area array pads on the PCB board. Examples are enhanced flexible CSP, chip-on-flex CSP (COF-CSP), fine pitch ball grid array (FPBGA), micro BGA, and memory-on-flex CSP MOF-CSP. COF-CSP and flex, PAC will be introduced here.

Based on the MCM-F (Multi-chip Module on Flex) technology of GE and Lockheed-Martin, COF-CSP is a kind of BGA CSP with a flexible middle support layer. It features via holes on flexible substrate made by laser drilling. The interconnection between the chip and the middle support layer is formed with a metallized layer by sputtering or electroplating. Usually, board level interconnection is realized by a eutectic BGA bump with a bump pitch of 0.5 mm.

COF-CSP was developed for packaging IC chips with a small to moderate number of I/O pins. Its potential applications include memory packages and ASIC in portable electronic appliances. The package size is quite compact, corresponding to the definition of CSP. With the mechanical thinning process of silicon chips, the thinnest package thickness can reach 0.25 mm. This package is currently being evaluated and reliability tested. Good electronic and thermal performance is expected. Customers from the communications industry now are evaluating its application in digital cell phones for packaging chip sets with general pad layout and common PCB coverage.

As shown in Figure 5.15, flexPAC is one kind of CSP based on a single piece of flexible substrate, which was developed by the Berlin University of Technology.[6] In this package, Ni/Au under ball metallization (UBM) is formed on the wafer by chemical plating, followed by the bumping process using eutectic Au/Sn materials. Chip-level interconnection is realized by penetrating flexible substrate with laser beams to heat up bumps in order to join bumps with the copper pads on flexible substrate. The solder balls as output pins of the package are connected with the copper circuit through the holes in the polyamide substrate. Typical bump diameter and pitch is 0.3 mm and 0.8 mm, respectively.

Metallization of flexPAC's bumps has a high performance-to-price ratio, and the production procedure may operate in roll-to-roll mode. All these advantages are beneficial to cost and mass production. The use of nontraditional packages, however, presents an obstacle to the initial investment of setting up new production equipment. Several qualifying tests showed that the moisture resistance of flexPAC packages meets the requirements of class 1 of JEDEC Standards, and those packages also have good reliability in thermal cycling and high-temperature storage. For its thin and compact structure, flexPAC can be applied in packaging memories and ASIC devices in portable equipment.

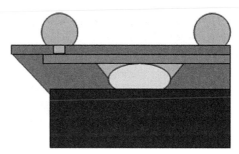

Figure 5.15   Schematic diagram of flexPAC

3. Rigid Substrate CSP

This kind of CSP makes use of a rigid substrate to redistribute fine pitch pads on a chip to form wider pitch area array pads on a PCB, for example, chip array packages, low cost bump flip chips, ceramic miniball array packages, plastic chip carrier (PCCs), and transformed grid arrays (TGAs). PCC and TGA will be discussed below.

PCC is a kind of leadless plastic package using a rigid middle support layer. This kind of CSP uses metallized organic substrate as a chip carrier by using common wire bonding as the first level interconnection. In place of molding plastic packages, it uses ball-shaped glue encapsulation by a top-dispenser to protect chips and bonded wires. The total thickness of the package including the encapsulation part is 1.0 mm. The second level interconnection uses electroplated pads with the smallest pad pitch 0.65 mm. The surface pads under the middle support layer are connected with the upper-surface metallized layer through through-holes. The through-holes are left unfilled for electronic connection through metallization. More through-holes can be punched under the chip attaching area to improve thermal performance. Therefore, PCC's thermal performance will be better than that of normal molding plastic package. A series of tests proved that this kind of CSP has the same class of reliability that other surface mount devices do.

Compared with common plastic packages with the same pad distribution, PCCs can hold bigger silicon chips. What's more, with the use of lead-free design, the installation area of the PCC is smaller than a lead package with the same size. Therefore, the shape factor is the primary advantage of the PCC package. Moreover, it is expected that the package would be a type of CSP with a relatively low cost, since PCC manufacturing is compatible with existing package infrastructure and technology. PCC package development is mainly for application in devices with fewer pin numbers, such as modules in wireless communication. It is estimated that CSP will replace TSOP in the near future.

Sony's TGA is a kind of CSP using rigid substrate. The package middle support layer is a laminated FR-4 (Flam Retardant) with electroplated through-holes filled with resin and then metallized to form pads on both sides of the support layer. In order to achieve chip-level and board-level connection, the pads are precoated with eutectic solder, which is a key process in TGA packaging.

TGA's first-level interconnection uses high-lead solder bumps. The redistribution of I/O can be done on the chip or middle support layer. To ensure the reliability, underfill adhesive is used to envelop solder joints. The second-level interconnection uses Land grid array (LGA) with A pin layout of quad flat array of 0.5 mm pitch. To deal with such a fine pitch structure, the board level assembling of TGA package uses an eight-layer stacking PCB. Results from a series of tests show that the TGA package has good package reliability and long fatigue life of the solder joints.

The advantages of the TGA package include simple structure, light thickness, and compact size. In addition, since the middle package support layer is organic substrate and its process is compatible with the existing infrastructure, the cost of such a CSP is relatively lower. TGA is an attractive solution for packaging ASICs of portable electronic products. TGA packages are used in Sony's Digital Camera VCR "DCR-PC7." With the adoption of this kind of CSP, the installation area of PCB was reduced by 37%. Furthermore, a new type of TGA with wire bonding technology had been developed, which could be applied in CPUs and DSP (digital signal processor) of cell phones.

4. Wafer-level Redistribution CSP

Wafer-level redistribution CSP makes use of metal layers on the wafer to redistribute finer pitch pads around the chip to form wider pitch area array pads on the PCB, such as micro SMT (MSMT), UltraCSP, SuperCSP, minitype solder ball array package, and Shell-PACK/Shell-BGA. UltraCSP will be introduced as follows:

Flip Chip International (FCI)'s UltraCSP is a kind of flip-chip package redistributing I/O on a wafer. As shown in Figure 5.16, the redistribution is composed of two layers of polymer dielectric and one UBM layer. The first polymer layer is directly coated on the passivation layer of the active part of the IC. Via holes are drilled in the passive layer on the top of the chip's pad, where the first-level interconnection is achieved through the UBM deposition on the holes. By sputtering the UBM patterning layer, the required printing wiring is formed. The second-level interconnection uses eutectic bumps or high-lead solder bumps, which are joined with UBM bonding pads restricted by the opening windows of the second polymer layer. The height and minimum pitch of bumps are about 0.25–0.4 mm and 0.4 mm, respectively.

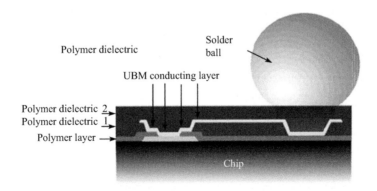

Figure 5.16   Schematic of ultraCSP

UltraCSP is a kind of true CSP with low thickness, less than 1.0 mm, including bump structure. It is mainly used in packaging ICs with a small to medium number of I/O pins of 150 μm pitch. The target application is memory modules (flash memory, SRAM (static Random Access Memory), DRAM). Since it can redistribute pads, UltraCSP can provide common PCB pads for IC devices from different manufacturers. In addition to I/O redistribution capabilities, another important advantage of UltraCSP is that the UBM is able to repair scratches on the chip bonding pads caused by IC testing. Thus, the technology is very suitable for packaging the wafer after being tested. Like other wafer level CSPs, UltraCSP is easy to adapt while the chip size decreases or the wafer size enlarges. Therefore, for UltraCSP packages, the reduction of chip size and enlargement of wafer size will significantly reduce the package cost of each IC chip.

5.6   Development Prospects

While single chips still cannot reach the scale of multichip integration, We propose that the highly–integrated, high performance, and highly–reliable CSP chips and special integrated circuit chips could be assembled into various electronic modules, subsystems, or systems on a high-density multilayer interconnected substrate with surface mount technology. MCM technology came out based on this prospect, and it has heavily influenced modern computers, automation, and communications.

Along with the development of LSI design and manufacturing techniques, as well as the application of nanotechnology, the chip size has been greatly reduced, and assembling several LSI chips into a package with a fine wiring multilayer substrate to form MCM product has been proposed. Further, another proposal is to integrate various circuit chips on a big wafer to a shift the package from a single small chip-level to a silicon wafer-level package. Based on this, the ideas of system-on-chip (SOC) and PC-on-chip (PCOC) were developed.

With the development of CPUs and other ULSI circuits, package types also developed correspondingly, which in turn promotes the development of chip technology; thus a close relationship of mutual promotion and mutual influence between them has developed.

Questions

(1) Please give the types of device-level package and their respective features. What are their main applications?

(2) List recent popular plastic package products.

(3) What is the future of CSP?

References

[1] Lihua Ning, Guilin Zhao, Yongsong Ye, et al. Reliability Research on Parallel Seam Welding Lids. Electronics and Packaging, 10(2005): 24–25, 48.

[2] Yufeng Jin, Jinwen Zhang, Yilong Hao, et al. A novel vacuum packaging for micromachined gyroscope by low temperature solder sealing. In: Proceedings of the Fourth International Symposium on Electronic Packaging Technology, 2001: 270–273.

[3] Jin Geng. Application Prospect of Ceramic Packing. Advanced Ceramics, 4(2004): 41–42.

[4] John H. Lau. Ball Grid Array Technology. New York: McGraw-Hill, 1995.

[5] John Lau, S. W. Ricky Lee. Chip Scale Package: Design, Materials, Process, Reliability, and Application, New York: McGraw-Hill, 1999.

[6] Kallmayer, C., Azadeh, R. Becker, K.-F., et al. A Low Cost Approach to CSP Based on Meniscus Bumping, Laser Bonding Through Flex and Laser Solder Ball Placement. In: First Electronic Packaging Technology Conference (EPTC), (1997): 34–40.

CHAPTER 6

MEMS Packaging

6.1 Introduction

Micro-Electro-Mechanical Systems (MEMS) is the integration of a number of microcomponents on a single chip using microfabrication technologies. The electronics, mechanical, and electromechanical components are fabricated using technologies borrowed heavily, but not exclusively, from integrated circuit fabrication technology. Typical dimensions of MEMS devices are less than 1 mm, with feature sizes on the order of microns. Scanning electron microscopy (SEM) pictures of some MEMS devices are shown in Figure 6.1.[1,2]

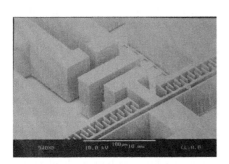

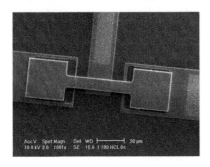

Figure 6.1   SEM of some MEMS devices

MEMS is an interdisciplinary field combining electrical engineering, mechanical engineering, and physics/chemistry; it is also referred to as micromachines, or micro systems technology (MST). MEMS promises to revolutionize nearly every product category by bringing together silicon-based microelectronics with other micromachining technologies, making possible the realization of complete systems-on-a-chip. MEMS is also an enabling technology allowing the development of smart products, augmenting the computational ability of microelectronics with the perception and control capabilities of microsensors and microactuators; applications include automotive accelerometers and pressure sensors, inkjet nozzles, fiber optic switching systems, biomedical instruments, etc.

MEMS packaging represents one of the greatest challenges our industry has ever faced, because the typical three-dimensional structures and moving elements of many MEMS devices generally require some sort of cavity package to provide free space above the active surface of the MEMS device. As a matter of fact, the cost, long-term reliability, and yield of many microsystems are often dictated by the package. Ideally, an advanced packaging technology for MEMS should provide a solution with low cost capable of being handled by existing facilities and technologies. In particular, it should protect MEMS devices from their operation environment. That is, it should provide MEMS devices and on-chip circuits with functions such as mechanical support, protection from the environment, electrical interconnection, and thermal management.

MEMS packaging is a critical factor in the integration and commercialization of microsystems, which is responsible for 75% to 95% of the overall cost. Closely tied with the IC silicon-processing technology, which is widely used currently, MEMS packaging can take advantage of these mature chip-scale packaging techniques, including flip-chip and ball-grid-array techniques. However, owing to its diversity, MEMS packaging is still complicated. Recently, developments in MEMS have led to growing interests in MEMS packaging at the wafer level to reduce the packaging and testing cost. Various approaches in this area can be characterized into two categories: thin film encapsulation and microcap using wafer bonding technologies. All along, reliability must also be considered. This chapter will summarize the primary package types that apply to MEMS technology and their specific concerns.

6.2   Function of MEMS Packaging

MEMS packages can contain many electrical and mechanical components. Packaging serves to integrate all of the components required for a system application in a manner that minimizes size, cost, mass, and complexity. The package provides the interface between the components and the overall system, as shown in Figure 6.2. A MEMS die sawed from a wafer alone is extremely fragile and must be protected from mechanical damage and hostile environments. To function, electrical circuits need to be supplied with electrical energy, which is consumed and transformed into mechanical and thermal energy. Therefore, the three main functions of a MEMS package include mechanical support, device protection, and electrical interconnections to other system components[3−5].

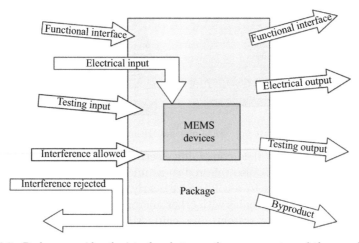

Figure 6.2   Package provides the interface between the components and the overall system

6.2.1   Mechanical Support

Owing to the very nature of MEMS being mechanical, the requirement to support and protect the device from thermal and mechanical shock, vibration, high acceleration, particles, and other physical damage during storage and operation of the device becomes critical. The mechanical stress endured depends on the application. For example, launching a missile generates greater mechanical shock than that experienced by consumer electronic devices.

The coefficient of thermal expansion (CTE) of the package should be equal to or slightly greater than the CTE of the substrate (silicon in most cases) for reliability, since thermal shock or thermal cycling may cause die cracking and delamination if the materials are unmatched or if the substrate is subject to tensile stress. Other important material properties are thermal resistance of the carrier, the material's electrical properties, and its chemical properties, or resistance to corrosion.

Once the MEMS device is supported on a (chip) carrier, the wire bonds or other electrical connections are made, and the whole assembly must be protected from particles and other physical damage. This is accomplished either by adding walls and a cover to the chip or by encapsulating the assembly in plastic or another material. Since the electrical connections to the package are usually made through the walls, the walls are typically made from glass or ceramic. The glass or ceramic can also be used to provide electrical insulation of the leads as they exit through a metal package wall. Although the CTE of the package walls and lid do not have to match the CTE of silicon-based MEMS, since they are not in intimate contact, it should match the CTE of the carrier to which they are connected.

6.2.2 Device Protection

1. Mechanical Packaging

The traditional "hermeticity" that is generally thought of for protecting microelectronic devices may not apply to all MEMS devices. Many MEMS devices are designed to sense physical/chemical/biological signals or even act as a driver. Therefore, nonelectronic access must be provided to exchange energy and material with the surrounding environment. These devices range from microphones and gas sensors to implanted biomedical sensors that operate in certain types of liquids. These devices might be directly mounted to a printed circuit board (PCB) or a hybrid-like ceramic substrate and have nothing but a "housing" to protect it from mechanical damage such as dropping. On the other hand, the package design of such devices can be very complicated, because it will greatly affect the device performance. It is advised that the package should be taken into consideration along with the device itself.

2. Hermetic Packaging

Many elements in the environment can cause corrosion or physical damage to the metal lines of a MEMS device as well as other components in the package. For example, moisture is a major failure mechanism in MEMS, the susceptibility of MEMS to moisture damage is dependent on the materials used: Al lines can corrode quickly in the presence of moisture, whereas Au lines degrade very slowly in moisture. Also, junctions of dissimilar metals can corrode in the presence of moisture. To minimize these failure mechanisms, MEMS packages for high reliability applications may need to be hermetic with the substrate, sidewalls, and lid constructed from materials that are good barriers to liquids and gases and do not trap gasses that are later released. Hermetic sealing can protect beam, contactor, and sensitive structure from moisture and other harmful gases and at same time protect MEMS devices from waste and liquid during profabrication, such as wafer dicing.

3. Vacuum Packaging

MEMS vacuum packaging, the technology for providing a gas-tight enclosure with an internal tiny cavity, has been a key technology for various MEMS devices and systems. Many electrical, mechanical, and optical sensors and actuators require vacuum packaging to preserve product integrity and safety. The reliability and performance of various MEMS devices, such as microresonator, thin-diaphragm pressure sensors, radio-frequency MEMS components, and some optical MEMS devices, are greatly affected by environmental pressure and composition. Vacuum packaging protects microdevices and systems from external environments and allows easy handling. The stability of pressure is very important to ensure dynamic performance, such as resonating frequency, of mobile parts.

Conventional vacuum packaging technologies have been around for over one hundred years, since the earliest phase of the electronics revolution, for example, the electric vacuum bulb. However, it is difficult to directly apply them to MEMS packaging in most cases. The challenge for MEMS vacuum packaging comes from the miniaturized structures and cavities. Some small change in materials properties and tiny leakages, which would not affect the macroscale devices, may dramatically alter the performance of MEMS devices.

The pressure deterioration within a microcavity is strongly related to the leakage of the packaging structure. Some sources of leakage include gas flowing from outside with higher pressure through a leaking path, permeation gas through the walls, and gases absorbed from inner surfaces. Therefore, design and fabrication should be undertaken carefully to achieve high quality packaging.

6.2.3   Electrical Interconnections

Because the package is the primary interface between MEMS and the system, it must be capable of transferring DC power and in some designs, RF signals. In addition, the package may be required to distribute the DC and RF power to other components inside the package. The drive to reduce costs and system size by integrating more MEMS and other components into a single package increases electrical distribution problems, since the number of interconnections within the package increases.

When designs also require high-frequency RF signals, the signals can be introduced into the package along metal lines passing through the package walls. The connection between a MEMS and DC and RF lines is usually made with wire bonding, although flip-chip die attachment and multilayer interconnects using thin dielectric components may also be possible.

6.3   Device-level Package for MEMS

Each MEMS application usually requires a new package design to optimize its performance or to meet the needs of a system. For single MEMS chip, typical integrated circuit standard packages or modified standard packages, like TO-packages, ceramic packages, or preformed injection molded packages can be used. The sealing techniques (soldering, welding, clogging, shedding) for these packages are compatible with the technology steps used for packaging of microelectronic devices. Three of the most common technologies—metal packages, ceramic packages, and plastic packages—are presented below.

6.3.1   Metal Packages

Metal packages are often used for microwave multichip modules and hybrid circuits because they provide excellent thermal dissipation and excellent electromagnetic shielding. They can have a large internal volume while still maintaining mechanical reliability. The package can use either an integrated base and sidewalls with a lid or it can have a separate base, sidewalls, and lid. Inside the package, ceramic substrates or chip carriers are required for use with the feedthroughs.

The selection of a proper metal can be critical. CuW (10/90), SilvarTM (a Ni-Fe alloy), CuMo (15/85), and CuW (15/85) all have good thermal conductivity and a higher CTE than that of silicon, which makes them good choices. KovarTM, a Fe-Ni-Co alloy, is commonly used. All of the above materials, in addition to Alloy-46, may be used for the sidewalls and lid. Cu, Ag, or Au plating of the packages is commonly done.

Before final assembly, baking is usually performed to drive out any trapped gas or moisture. This reduces the onset of corrosion-related failures. During assembly, the highest temperature curing epoxies or solders should be used first, and subsequent processing temperatures should decrease until the final lid seal is done at the lowest temperature to avoid the processes at later steps damaging the parts formed in the earlier steps. Au-Sn is a commonly used solder that works well when the two materials to be joined have similar CTEs. Figure 6.3 is the metal packaged MEMS pressure sensor.

Figure 6.3   A metal packaged MEMS sensor

6.3.2   Ceramic Packaging

Ceramic packages have several advantages that make them especially useful for microelectronics as well as MEMS. They provide low mass, are easily mass produced, and can be low in cost. They can be made hermetic, and can more easily integrate signal distribution lines and feedthroughs. They can be machined to perform many different functions (seen in Figure 6.4). By incorporating multiple layers of ceramics and interconnect lines, the electrical performance of a package can be tailored to meet design requirements. These multilayer packages offer significant size and mass reduction over metal-walled packages. Most of that advantage is derived by the use of three dimensions instead of two for interconnect lines.

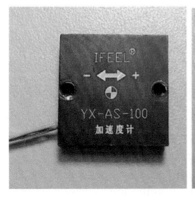
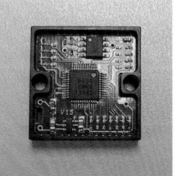

Figure 6.4   A ceramic packaged MEMS accelerometer (Courtesy of FirstMEMS Inc.)

6.3.3   Plastic Packaging

Plastic packages have been widely used by the electronics industry for many years and for almost every application because of their low manufacturing cost (seen in Figure 6.5). High-reliability applications are an exception because serious reliability questions have been raised. Unlike the metal and ceramic packages, plastic packages are not hermetic, and hermetic seals are generally required for high-reliability applications. The packages are also susceptible to cracking in humid environments during temperature cycling of the surface mount assembly of the package to a motherboard. However, plastic packaging is a well-proven technology and is very cost efficient.

Figure 6.5   A plastic packaged MEMS pressure sensor (Courtesy of FirstMEMS Inc.)

6.4   Wafer-level Package for MEMS[6−9]

Although some conventional IC techniques can be used for chip-level MEMS packages, fragile MEMS structures and delicate membranes must be protected from damaging processes such as dicing and cleaning; furthermore, movable MEMS devices must be packaged immediately after the mechanical elements are released, to prevent particles and stiction. The release and package should be done at the wafer level to reduce the cost, as shown in Figure 6.6.

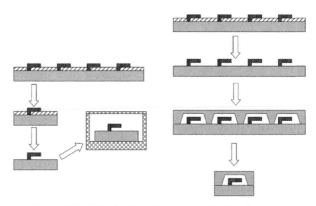

Figure 6.6   Chip-level package vs. wafer-level package

There are two methods to package a MEMS device on the wafer scale: either by bonding a cap on top of it, or by integrating the package into the process flow, as shown in Figure 6.7. Wafer bonding is being investigated, and thin-film approaches are being increasingly developed. No matter which kind of technology is used to package freestanding MEMS structure, the packaging process should be considered from the beginning of the system development, and it should be integrated into the device fabrication process.

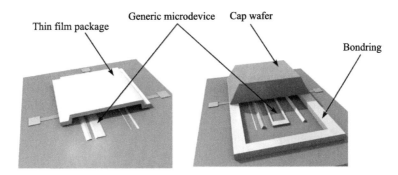

Figure 6.7　Thin film package vs. microcap package

6.5 Sealing Techniques

6.5.1 Introduction[10]

To prevent thermal and mechanical shock, vibration, high acceleration, particles, and other physical damage during storage and operation, most MEMS devices must be hermetically sealed, usually by bonding a microcap. A number of techniques for bonding and hermetically sealing silicon wafers at elevated temperatures have been demonstrated.

The wafer bonding process uses different bonding methods like anodic bonding, fusion bonding, and eutectic bonding to encapsulate microstructures by using a second substrate of silicon, glass, or other materials. This is a technology that has found widespread use in IC as well as MEMS fabrication. Applications include packaging, fabrication of 3-D structure, and multilayer devices. As for packaging, this process is about bonding the surfaces of wafers to serve as a hermetic seal of the microdevice. This process can bring the MEMS packaging to the wafer-level.

6.5.2 Eutectic Bonding and Solder Bonding

One commonly used wafer-bonding method is based on solder, or called eutectic bonding. In its simplest form, solder of a suitable material set can be formed in the bonding area between substrates of the package and the device. The two substrates are brought together and the temperature is raised until the solder flows and creates a bond. The most obvious materials to use are those standard solders used in microelectronic applications. However, the disadvantage of many such solder materials is that they contain either flux or sufficient impurities, which generate significant outgassing during the reflow process. This creates a major problem when trying to use such solders for vacuum packaging. Instead of standard solder, it is also possible to use alloys of different materials in the form of eutectic solder. One of the most common material sets is the eutectic of gold and silicon, which is a new fluxless solder material.

Silicon-gold eutectic is quite attractive because it is formed at a temperature of 363°C with one part silicon and four parts gold. These materials are commonly used in MEMS fabrication, and when the eutectic is formed, outgassing is not a problem since the mixture is simply formed by raising the temperature and the starting materials are pure. In addition, the temperature is low enough for most applications.

6.5.3   Adhesive Bonding

The advantage of adhesive bonding is it has a low process temperature and it makes joining different materials possible. Adhesive bonding is used to join two substrate materials with an intermediate adhesive layer, such as epoxies or polymers. Sometimes epoxy is acceptable for a gas-filled MEMS device, such as in a micro-optical switch, for holding optical components together. However, epoxy in the light path is not desirable since it may age, drift, or crack at high laser power levels. This creates a significant problem for the package, since the package has to protect the device and simultaneously provide access to the environment that the device is supposed to interact with. As a result, a lot of effort has been put into developing the proper protection/encapsulation medium for MEMS.

Figure 6.8 is an example of application of adhesive bonding for a micro-optical switch. The process of adhesive bonding starts with applying the adhesive layer, followed by contacting the wafers and forming bonds by a heat curing, or ultraviolet (UV) curing.

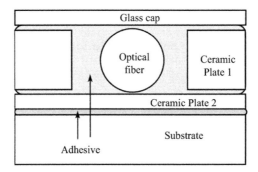

Figure 6.8   Adhesive packaging for micro-optical switch

Adhesives are widely used in packaging for MOEMS, such as tacking, filling, and sealing the precision structure, joining the ceramic frames, glass lid, and PCB substrate to form a hermetic package.

However, it is difficult to obtain uniform and hermetic bonding with vacuum grade that is not sensitive to humidity due to the permeation of moisture. The solution could be to choose an adhesive material with a low permeation rate or coat an antipermeation layer such as SiO_2.

Developments in adhesive polymer bonding have opened up new options to join wafers that are not planar or not feasible to planarize before bonding, to enable wafer-scale packaging. Adhesive bonding offers several potential advantages:

(1) It has a low bonding temperature: bonding temperature can be below 120°C, depending on the adhesive material, and thus most adhesive bonding processes are compatible with CMOS circuits.

(2) Various wafer substrate materials can be used.

(3) It is low in cost.

(4) Wafers with structured surfaces can be bonded as long as their dimensions are lower than the thickness of the adhesive.

(5) It has a high bonding strength.

(6) It is compatibe with standard cleanroom processing.

However, this technology has its limitations: polymers are inherently permeable to gas and moisture and not suitable for hermetic seals. In previous publications concerning adhesive bonding[3−5], it was concluded that adhesive bonding is not suitable for packaging applications that lead to hermetic sealing.

6.5.4   Glass Frit Bonding

The advantage of glass frit bonding is that it is capabile of producing good hermetic seals. The glass frit bonding process was developed to create an in-between glass layer at temperatures below 400°C. By combining anodic bonding with glass frit coating on wafers in various materials, such as silicon, ceramic, and metal, it is possible to anodically bond wafers other than glass wafers with silicon wafers. In addition, it can be used in hermetic bonding between ceramic layers.

The process can be described as below. After applying the frit paste onto the substrate with MEMS chips through screen printing process, the frit must be thoroughly dried. Oven drying can be used. The temperature is then raised to around 400°C, the softening point of the frit, and held for 5 minutes to 10 minutes before cooling down. Sealing cycles depends on the geometry and size of sealing interface. The important parameters of the heating process include starting point for experimentation, sealing temperature, holding temperature, and heating rate of each step, which should follow the specifications given by the frit supplier. Maintaining an oxidizing environment in the furnace at all times is necessary.

6.5.5   Anodic Bonding

Nowadays anodic bonding is widely applied in vacuum packaging of MEMS devices. It is a reliable and effective process for hermetically sealing a silicon wafer to a glass wafer or a quartz substrate. Anodic bonding is usually performed under constant temperature and voltage. The cathode makes contact with the glass, while the anode connects to the silicon wafer. By heating at 200–500°C and applying 200–1500 Volts DC across a silicon-glass wafer stack, the positive ions in the glass, mainly sodium ions coming from the dissociation of Na-O, migrate toward the cathode, leaving behind the nonbridging oxygen ions. Consequently, a negatively charged depletion layer is formed adjacent to the anode. The electrostatic force between this negative layer and the positive charge induced on the anode brings the two sides into intimate contact. This force, allied to the softening of the glass, allows some conforming of glass to the opposing surface and makes possible hermetic bonding between surfaces that are imperfect. Figure 6.9 shows an anodically bonded Si and glass stack.

Figure 6.9   An anodically bonded Si and glass stack

6.5.6   Fusion Bonding

Fusion bonding is another common wafer bonding technique used in MEMS packaging. A typical application of this technique is silicon-silicon bonding. Unlike anodic bonding, silicon fusion bonding relies on chemical force rather than electric force for bonding. To achieve bonding, the boding surface of each wafer must be treated with a hydration process, to introduce oxygen-hydrogen (O-H) bonds to the interface. This process can be done by soaking silicon wafers in HNO_3 or H_2O_2-H_2SO_4 solvent. It is also important that the bonding surfaces should be extremely flat and particulate free. After surface treatment, two wafers are brought into contact and some pressure is applied to make them stick initially. Then a high-temperature annealing is performed at about 1000°C to create a strong Si-O-Si bond through a dehydration process.

Silicon fusion bonding is a simple technique and can produce very strong bonding. This technique is therefore used for the fabrication of high-pressure silicon sensors with low-cost packaging and high-pressure bipropellant for rocket engines. However, it has some strict requirements on the flatness and cleanness of the wafer surfaces. And, a high annealing temperature sometimes is not compatible with the fabrication process of a microsystem that contains electronic devices. To overcome this disadvantage, a new packaging process that combines the silicon fusion process and a localized heating technique has been successfully demonstrated. Microheaters, which are made of polysilicon, are patterned in the confined bonding region, to provide localized high-temperature heating for fusion bonding. Without regular global heating, the substrate temperature remains low.

6.6   Thin-film Encapsulation[6]

This process integrates the MEMS encapsulation steps with the device fabrication process. As a postprocessing step in the manufacturing flow, it is a low-cost technique for increasing yield and is more widely used today in the MEMS industry. In certain applications, thin-film encapsulation is sufficient as the final packaging of the device prior to use. This process mainly involves surface micromachining technologies, such as sacrificial layer deposition, etching, and thin film. Typical steps of the encapsulation process are illustrated in Figure 6.10. After the MEMS device is fabricated, instead of removing the sacrificial layer for the

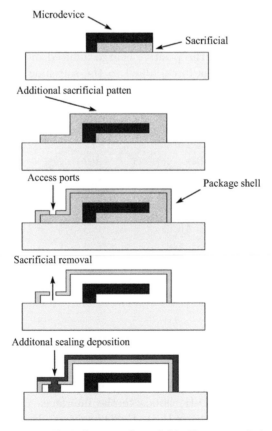

Figure 6.10   Typical process flow of thin-film encapsulation

device, another thicker layer of the same sacrificial material is deposited and patterned. Then the package cap layer is deposited and patterned to form the thin-film package shell with fluidic access ports for release etching. After removal of the sacrificial layer to release both the device and the package, the access ports will be sealed with another thin-film deposition.

Based on this process, there have been many thin-film encapsulation packages, fabricated with different packaging materials. Many different thin-film deposition methods, such as low pressure Chemical vapor deposition (LPCVD), physical vapor deposition (PVD), and electroplating, have been involved based on microsystem fabrication specifications. Lin et al. demonstrated a low-stress nitride thin-film package for micro-comb-drive resonator, using a process very similar to that illustrated in Figure 6.12. Phosphorus silicate glass (PSG) was selected as sacrificial material, which was removed by 49% hydrofluoric acid (HF) later. To accomplish the vacuum package for resonator application, LPCVD was used to deposit nitride film to form a package shell and seal the etchant access ports. The resulting package vacuum was about 200 mTorr.

6.7   Vacuum Packaging for MEMS[11,12]

Vacuum packaging is very critical for some MEMS devices to perform their basic functions properly and to enhance their reliability by keeping these devices away from harmful external environments. The mechanical quality factor of resonant MEMS devices, such as microresonators and microgyroscopes, will deteriorate with increased of environmental pressure due to the air damping effect.

The product life of a microvacuum field emitter device (FED) can be prolonged by hermetic sealing. In contrast, humidity and corrosive gases will degrade the reliability of microstructures due to the delaminating of thin films consisting of functional parts, the accumulation of surface charge in dielectric surfaces, and the stiction of mechanical structures caused by capillary forces.

Significant efforts on hermetic sealing and vacuum maintenance in MEMS packaging have been made. For instance, packages using capsules or shells were introduced to create a microvacuum cavity. A pressure of about 10^{-3} Torr inside vacuum-packaged microstructures has been reported by using SiO_2 film as the sealing material and applying anodic bonding with the glass cap and MEMS wafer. In addition to hermetic sealing, long-term maintenance is another necessity for vacuum packaging. In order to maintain a high vacuum in a cavity, residual gases must be eliminated. The sources of the residual gases include outgassing from the sealed structures, permeation through the package, and gas desorption from the inner surface of MEMS structures during the packaging process.

Gas permeation should be carefully considered when choosing materials used in MEMS packaging. Because of the smaller volume in the MEMS cavity, the pressure deterioration caused by gas permeation in MEMS is much more than those in conventional structures for the same quantity of permeated gas. Furthermore, thinner structures are often used in MEMS vacuum packaging. This will cause more serious permeation problems for MEMS devices. For instance, permeated gas increases one hundred times when the thickness of a wall or diaphragm is reduced from 1 mm to 10 μm.[1]

6.7.1   Vacuum Maintenance for MEMS Packaging[2]

With the advantage of high sorbing capability, commercial nonevaporable getters (NEGs) have been used in vacuum maintenance of electronic packaging. It is prepared by coating getter materials on strips and sheets and cutting into the desired shape and size by mechanical cutting or by laser beam. Then the NEG is fastened on the inner surface of the device's structure. However, it is difficult to use the commercial NEG for maintaining a higher vacuum environment in a microscale cavity in order match with the miniaturization of MEMS devices. The deposition of thin-film or thick-film of getter materials onto the inner surface of microstructures is a solution to maintain a vacuum in a microcavity.

A schematic on the key process steps is presented in Figure 6.11. The preprocess consists of mask design, making getter paste by mixing K_4Si and graphite with powder of Zr-V-Fe alloy, and the fabrication of the MEMS chip. The coating of NEG thick film starts with getter paste printing on the surface of double-side polished Pyrex 7740 glass wafer to form a pattern. After prebaking at 120°C for half an hour, the glass wafer and silicon wafer with MEMS structure are cleaned to eliminate particles and other contamination on the surfaces. Anodic bonding was then applied to hermetically join the glass wafer to the silicon wafer. The bonding process was performed with an EV 501 bonder at a pressure of 1×10^{-3} Torr and DC voltage of 1000 Volts for 60 minutes. The bonding temperature is about 450°C.

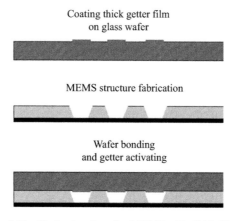

Figure 6.11   Packaging flow for MEMS with thick-film NEG

The sorption capability was then tested to examine the performance of the getter film. Experimental pressure variation against time is shown in Figure 6.12. Good sorption capability of 4.88×10^6 Pa·L/m² has been measured with the getter measuring 6.5 Pa·L/s.

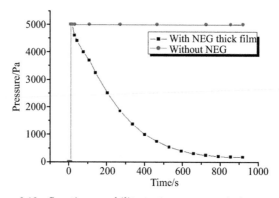

Figure 6.12   Sorption capability test: pressure variation vs. time

Flashing getter material has also been applied in MEMS packaging research due to its attractive features like steady performance, consistent yield of getter materials, and minimal outgassing during evaporation. It can be easily deposited onto the inner walls of the microcavity by evaporation in the form of thin film. The thickness of getter film coated on the wafer can be controlled in the range of several to hundreds of microns by adjusting the heater temperature and process time. It is also feasible to form a patterned getter film on the lid surface using a physical mask between the getter source and the target.

6.7.2   Leakage Detection

It is obvious that in the event of a leakage, the performance of vacuum packaged microsystems will deteriorate much more than that of a macroscale device does. More sensitive methodologies and measurement tools are required detect the suspected leakage in microscale cavities. Helium leak detection is one increasingly used method. Pressurized helium is introduced to enhance the He concentration within the tested component. A leakage rate as low as 5.8×10^{-10} Pa·m^3/s in a microvalve can be determined precisely from the pressure changes. An attached spinning rotor gauge (SRG) is another approach. The leakage rate of MEMS devices can also be monitored using embedded microsensors, such as field-emission transistors, micro capacitors, radiators, and resonator.

6.8   Case Study: A 3D Wafer-level Hermetic Packaging for MEMS[13]

In this session, a three-dimensional wafer-level hermetic packaging solution for MEMS is presented. We will use a pressure sensor as an example of a device to be packaged. The packaging technologies involve wafer thinning, wet etching, wafer bonding, vacuum sealing, 3D electric interconnection and postprocess for flip-chip device processes.

6.8.1   Design and Fabrication Process

The proposed 3D wafer-level approach for MEMS packaging is shown in Figure 6.13. Basically, the packaging structure is composed of three stacked wafers. The MEMS wafer is sandwiched between a top glass wafer and a bottom substrate wafer.

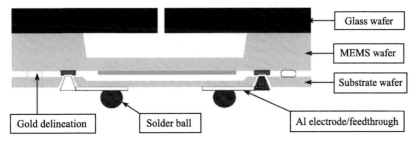

Figure 6.13   A 3D wafer-level approach for MEMS packaging

The bottom Si wafer or substrate wafer with a cavity provides hermetic sealing with one side of the MEMS structure, along with a 3D electric feed-through connecting the metal pads on the MEMS wafer with solder balls or bonding pads for the flip-chip process. The top one, a perforated glass cap wafer, protects another side of the MEMS structure during the packaging process and in some applications. Opening a hole on the top wafer will establish a physical connection between the environment and the MEMS sensor.

The fabrication process flow started with wafer preparation, including perforated glass cap wafer, MEMS wafer, and Si substrate wafer with wet etched via holes (a). After gold delineation on the MEMS wafer was fabricated (b), the MEMS wafer and substrate wafer were hermetically bonded by eutectic bonding (c). Then Al metal film for vacuum sealing and seed layer was sputtered (d), followed by glass top wafer anodically bonded to the MEMS wafer (e). Finally, the postprocess, which includes Al layer patterning (g), Cr/Ni/Au UBM layer lithography and patterning (f), and solder bumping (g), was carried out.

6.8.2   Experiments and Results

1. Wafer Preparation

A MEMS wafer was prepared by typical micromachining fabrication. A silicon-on-insulator (SOI) wafer was used to fabricate pressure sensors. The size of each sensing unit was (2.6 × 2.6) mm². The membrane with sensing piezoresistor was fabricated using the doping and etching process. The metal pads for electrical control and signal output were formed by deposition of aluminum film. The pressure sensor works from 1 bar to 10 bars. The top wafer was a 4-inch Pyrex 7740 perforated borosilicate glass wafer. With the chemical-mechanical polishing (CMP) process, it was thinned to 200 µm before it was anodically bonded to the MEMS wafer.

The substrate wafer used in this study was a 4-inch (100) standard bare wafer. After thinning to a thickness of 200 µm, a wet etching process was carried out to form via holes. A cavity was also etched to provide a space for sensor movement and a chamber of reference pressure.

2. Wafer thinning

In order to reduce the overall size of the stacking packaging and improve the quality of electric interconnection between metal pads and solder balls, a thinned wafer was used as the substrate wafer. A cost-effective wafer thinning process, face grinding, was developed. The reduction rate of the wafer thickness achieved 90 µm/min with 60 µm diamond abrasive, which can be effectively used to thin silicon wafers rapidly. It was noted that grinding with 60 µm abrasive at a load of 150 N generated scratches with a maximum depth no greater than 8 µm.

Experiments proved that an interface with nanoscale roughness on the wafer to be directly bonded is required to achieve high bonding quality. In general, Ra is recommended to 1 nm or less for wafer bonding. Therefore, mechanical polishing was performed with 9 µm diamond slurry to minimize the roughness of the wafer. Subsequently, CMP was carried out to further reduce the damage to a negligible level. The surface roughness of the thinned Si wafer was measured with an atomic force microscope (AFM). The parameters Ra and Rt were used to characterize the roughness of the targets. After the CMP process, the values of Ra and Rt achieved in this experiment were 0.649 nm and 3.578 nm, respectively, which is within the requirements of wafer bonding.

3. Hermetic Wafer Bonding

Eutectic bonding: For silicon-gold eutectic bonding, although the eutectic point is 363°C, the bonding temperature must be higher. A higher temperature can promote the diffusion of gold and silicon into each other and increase the thickness of the diffusion layer where the chemical composition can match what is needed for eutectic bonding. Hermetic inspection results indicated that the cavities were well sealed. The pull test results showed that the bonding strength was more than 5 MPa.

The patterned gold film on the MEMS wafer was obtained using the lift-off process. After spinning on a layer of photoresist, it was UV exposed and developed. With the lift-off process, the gold film with a width of 0.2 mm at the outer edges was left. To achieve a good bonding, a thicker gold layer was built up by the plating method.

Prior to eutectic bonding, the substrate wafer was dipped in HF solution for 10 seconds to remove surface contaminant and native oxide. After alignment, the stacked sensor and cap wafers were put in an EV510 bonder to perform eutectic bonding. After eutectic bonding, an aluminum layer with a thickness of 1 µm was sputtered on the silicon cap wafer. The actual pressure of the sputtering chamber is 0.3 Pa. During the sputtering process, the metal pads were built up and connected with metallized via holes. Finally, the cavities providing space for membrane movement were hermetically sealed.

4. 3D Interconnection

A vertical via electrode method for 3-D interconnection to realize hermetic sealing of silicon to silicon was developed for the package. A via hole was wet-etched on a Si substrate wafer, followed by eutectic bonding to the MEMS wafer prepared by standard MEMS processes. The vertical electrodes were formed through the vias by deposition and patterning of the metal film.

(1) Silicon Etching.

Anisotropic etching of single-crystal silicon with KOH solution was an easy way to create via holes on a Si wafer. Silicon dioxide and silicon nitride are the main etching mask materials, chosen for their slower etching rate than silicon. A silicon nitride film with thickness of 200 nm and a silicon oxide film with thickness of 100 nm were sequentially deposited on a silicon wafer by the plasma-enhanced chemical vapor deposition method. With the protection of photoresist mask, the Si_3N_4/SiO_2 layer was selectively etched by plasma dry etching.

It is noted that the etching rate increases with increasing KOH concentration and reaches a maximum of 1.45 μm/min in the range of 10wt% to 20wt% KOH. However, such high etching rates are achieved at the expense of good surface quality. The surface roughness is lowest when the etching solution contains 30wt% of KOH. Hence, the etching of silicon with 30wt% of KOH at a temperature of 80°C yields a reasonably high etching rate (1.3 μm/min) with good surface quality of the etched silicon.

(2) Fabrication of electrical via.

An aluminum layer with thickness of 1 μm was sputtered on silicon cap wafer using a DC magnetron sputtering system. The film was thick enough to fill the small gap between the MEMS wafer and the substrate wafer, so that the metal pads on the MEMS wafer were brought in to connection with the metallized via hole side walls as well as the subsequently delineated rerouting pads. At the same time, the metal vias, with their low permeation rate to moisture and gas molecules, also function as "plugs" in the sealing process by filling metallic material into the vias to form a hermetic seal. A typical electrical via interconnection is illustrated in Figure 6.14.

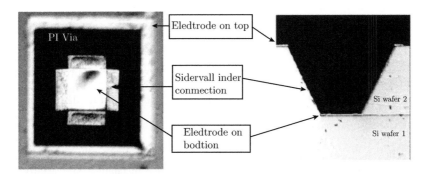

Figure 6.14   A typical electrical via interconnection (Courtesy of SIMTech)

5. Postprocess

The aluminum layer with a thickness of 1 μm was chosen as the seed layer due to its good adhesion with silicon. After patterning of the Al film, PVD sputtering was performed to deposit three layers of metals, Cr, Ni, and Au. Their thicknesses are 0.1 μm, 1.0 μm, and

0.1 μm, respectively. Cr acted as the adhesion layer, nickel as the UBM contact to solder and gold as a protection for nickel. Then, the patterned Cr/Ni/Au layer was used for solder bumping.

Three more layers of patterned films, Cr/Ni/Au, were deposited onto the substrate wafer, which served as a UBM layer for solder bumping. The standard photolithography process often resulted in open circuits of the electrical rerouting at the edges of the via holes. A photoresist that was normally applied for thick coating applications, such as bump electro-plating in wafer bumping, was used here. The photoresist was applied very slowly to the wafer, allowing it to flow into the holes to avoid air bubbles being trapped inside the holes. Finally, benzocyclobutene (BCB) was used to planarize the wafer surface and opened for solder bumping. A resultant pattern is shown in Figure 6.15.

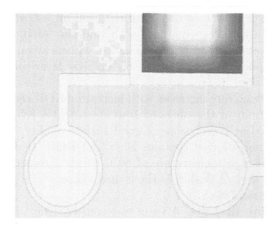

Figure 6.15   Aluminum rerouting and Al/Cr/Ni/Au UBM pads for solder bumping
(Courtesy of SIMTech)

To characterize the hermeticity of the packaged structures, the leakage tests were performed in a portable leak detector attached a bombing chamber with a sensitivity of 2×10^{-10} std cc/sec. No detectable leakage was found that proved the eutectic bonding and anodic bonding for the packaging resulted in very good vacuum sealing.

Questions

(1) What are the main features of MEMS packaging?

(2) What do bonding processes for MEMS packaging include? Please list their advantages and disadvantages.

(3) Why was wafer-level packaging widely applied in packaging for MEMS?

References

[1] Douglas Sparks, Jacob Trevino, Sonbol Massoud-Ansari, et al. An all-glass chip-scale MEUS package with variable cavity pressare. Journal of Micromechanics and Mioroengineering. 16(2006): 2488–2491.

[2] R.D. Gerke. Chapter 8: MEMS Packaging, it's not abook, But an onlime documentation. JPL PUB 99-1H.

[3] A. Jourdain, P. De Moor, K. Baert, et al. Mechanical and electrical characterization of BCB as a bond and seal material for cavities housing (RF-)MEMS devices. Journal of Micromechanics and Microengineering, 15(2005): S89–S96.

[4] Yexian Wu, Guanrong Tang, Jing Chen. A Low Temperature, Non-aggressive Wafer Level Hermetic Package with UV Cured SU8 Bond. In: Proceedings of MicroNanoChina07, MNC2007–21528.

[5] Ki-Il Kim, Jung-Mu Kim, Jong-Man Kim, et al. Packaging for RF MEMS devices using LTCC substrate and BCB adhesive layer. Journal of Micromechanics and Mioroengineering, 16(2006): 150–156.

[6] Xiangwei Zhu and Dean M. Aslam. CVD diamond thin lm technology for MEMS packaging. Diamond and Related Materials, 15(2006): 254–258.

[7] Hsueh-An Yang, Mingching Wu, Weileun Fang. Localized induction heating solder bonding for wafer level MEMS packaging. Journal of Micromechanics and Mioroengineering. 15(2005): 394–399.

[8] Xu Ji, Jing Chen, Ying Wang, et al. A wafer level hermetic package for micromachined structures on glass. In: Asia-Paci c Conference of Transducers and Micro-Nano Technology—APCOT 2006.

[9] Pejman Monajemi, Paul J. Joseph, Paul A. Kohl, et al. Wafer-level MEMS packaging via thermally released metal-organic membranes. Journal of Micromechanics and Mioroengineering, 16(2006): 742–750.

[10] Y. Jin, Wang, Z.F., Lim, P.C., et al. MEMS vacuum packaging technology and applications. In: Electronics Packaging Technology, 2003 5th Conference (EPTC 2003, Singapore), 10–12 Dec. 2003: 301–306.

[11] Y. Jin, J. Wei, P.C. Lim,. Z.F. Wang. Hermetic packaging of MEMS with thick electrodes by Si-Glass anodic bonding. Int. J. Comp. Eng. Sci., 4, 3(2003): 335–338.

[12] Yufeng Jin, Zhenfeng Wang, Lei Zhao, et al. Zr/V/Fe Thick Film for Vacuum packaging of MEMS. Journal of Micromechanics and Mioroengineering, 14, 5, (May 2004): 687–692.

[13] Y.F. Jin, J. Wei, G.J. Qi, et al. A 3-D Wafer Level Hermetical Packaging for MEMS. In: International Conference on Solid-State and Integrated Circuits Technology Proceedings, Oct. 18–21, 2004, Beijing, China, pp. 607–610.

CHAPTER 7

Module Assembly and Optoelectronic Packaging

7.1 Introduction

The module assembly is a package of one or several components on one substrate to form a functional block. The module assembly technology advances with component fabrication technologies, while it also influences the component packaging trend. A typical component package is DIP, as shown in Figure 7.1, where the module assembly is using through-hole technologies (THTs) in which holes are drilled in the PCB for component plug in and then components are fixed by wave soldering. Surface mount technology gained popularity in the 1980s, and a typical example is the quad flat pack, as shown in Figure 7.2. In the early 1990s, BGA increased SMT applications in high-density assemblies, as shown in Figure 7.3. The later developed CSP and WLP further expanded SMT usage in miniature and high-density module assemblies.[1,2]

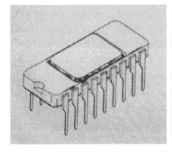
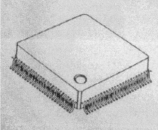

Figure 7.1   Dual in-line package (DIP)   Figure 7.2   Quad flat pack (QFP)   Figure 7.3   Ball grid array (BGA)

Optoelectronic packaging covers many areas, such as optoelectronic module and component packaging, where optoelectronic component packaging is the basic model. Optoelectronic components include active and passive components. Active components include light source, optoelectronic sensor, amplifier, and related modules, which includes light emitting diode, laser diode, photo diode, optical fiber amplifier, semiconductor laser amplifier, optical fiber Raman amplifier, etc. Passive components include connector, optical coupler, optical attenuator, optical isolator, optical switch, wavelength division multiplexing (WDM), optical fiber cable, etc. In addition, optical electronic integrated circuit (OEIC) and photonic integrated circuit (PIC) are involved in optoelectronic packaging. Owing to its complexity and diversity of optoelectronic devices, the related packaging technologies vary accordingly.

With a huge market, flat panel display packaging is unique in optoelectronic module packaging and involves special techniques such as anisotropic conductive film (ACF) in chip on glass (COG), transparent conductive indium tin oxide (ITO), mechanical connection, etc.

7.2   Surface Mount Technology

SMT is an electronic assembly technology that uses an automatic machine to assemble surface mount components onto a printed circuit board or other substrates directly. The components are usually discrete, small, with short leads or leadless. SMT is the critical technology for assembling discrete components into modules or devices.[1–6]

Unlike traditional THT, holes are not required in PCBs to solder components to the designated surface area. The SMT comprises the dispensing, screen printing, attachment, reflow, cleaning, and online testing processes. In brief, it uses tools to dispense glue or paste on the metal traces of a substrate, attaches surface mount devices (SMDs) to the correct locations, and builds electrical and mechanical connections after the reflow, as shown in Figure 7.4.

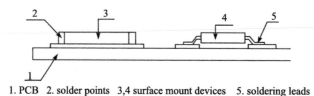

1. PCB   2. solder points   3,4 surface mount devices   5. soldering leads

Figure 7.4   A typical SMT assembly

SMT is the driving force in electronic devices, miniaturization, cost reduction, and increasing reliability, and is the milestone for the information technology industry. SMT encompasses SMD, attachment technology, and pick-and-place technologies. The high packing density of SMT enables the electronic product or system to achieve 40%–60% reduction in volume, a 60%–80% reduction in weight, and a 30%–50% reduction in cost. Along with excellent reliability and high-frequency performance of SMD, the SMT process and equipment selection and configuration are vital for the quality assurance of electronic products and systems.

Almost all of the electronic systems, telecommunications and computer networks in particular, have adopted SMT technologies. The world production of SMDs grows each year, and OEM and electronic manufacture service (EMS) have become mainstream in the electronics industry. Annual production of traditional DIPs, through-hole resistors, and capacitors are decreasing. SMT will dominate the market in the years to come.

Surface mount technologies involve two areas: one is the material, which includes components and its manufacture technologies along with supporting materials such as flux, epoxy, and solder; the other is assembly processes, which include screen printing, pick-and-place, reflow, cleaning, inspection, etc.

7.2.1   Features

The basic difference between SMT and THT lies in the operation of "placement" and "insertion," which determines many aspects of SMD, such as its packaging format, processes, equipment, and functional differences. Surface mount placement methods can be classified as single-side hybrid, double-side surface mount, or all surface mount. The features of SMT include the following:

(1) high component count and high quality requirement for SMD;

(2) high accuracy assembly and high yield;

(3) complicated assembly steps;

(4) highly automated, requiring automatic equipments;

(5) requiring technology and know-how.

1. The Benefits of SMT

(1) Since there is no through hole and wiring is via buried layers, there are more layout spaces and higher wiring density. With the same put in functionality, reducing the layer count lowers the overall cost.

(2) It has lower weight, which is suitable for highly mobile and light electronics, such as aviation, space, portable electronics, etc. Lighter weight also contributes to higher physical properties such as resistance to vibration.

(3) Its assembly speed is faster than through-hole, and it is easy to automate, up to 50 thousand parts per hour. This raises productivity and lowers assembly cost.

(4) New solder paste and reflow technologies have promoted the soldering quality and avoided short, cold soldering and deformation problems.

(5) It has a smaller footprint and shorter wiring; lower parasitic inductance and capacitance enable higher signal transmission speed, reduced noise, etc.

2. The Disadvantages of SMT

(1) SMT poses increased testing difficulty and costs more because of higher packaging density.

(2) SMT requires adaptation to traditional through-hole components to SMD types, such as chip resistor, chip capacitor, surface mountable quartz, transformer, switch and relay, etc.

7.2.2 Basic Assembly Processes and Work Flow

1. Assembly Method

There are three major SMT assemblies: single-side hybrid, double-side surface mount, and all surface mount. With minor differences in assembly methodology and processes, they share the following common work flow: inspection→screen printing of paste→pick and place→reflow→cleaning→inspection→rework.

The main processes are outlined in Figure 7.5.

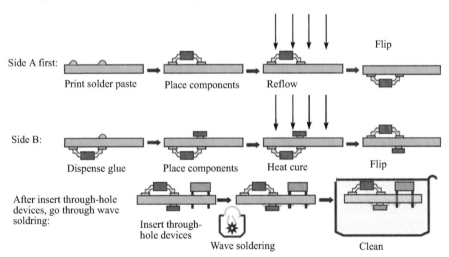

Figure 7.5　Hybrid assembly processes

2. Basic processes

(1) During screen printing the solder paste or glue is printed on the PCB and prepared for the soldering. The required equipment is a screen printer, which is an advanced machine in SMT production.

(2) During dispensing the glue is dispensed onto designated locations in order to fix components on the PCB. The required equipment for dispensing is a dispenser, which is normally in the front end of SMT production or right after inspection.

(3) During pick and place the SMD is picked up and placed on the correct location on

the PCB. The required equipment is a "pick and place" machine, which is located after the screen printer.

(4) During curing the glue is cured and holds components onto the PCB. The required equipment is a curing oven, which is located after "pick and place" machine.

(5) During reflow soldering the solder is melted and attachs the SMDs onto the PCB. The equipment required is a reflow oven, which is used after the "pick and place" machine in the SMT line.

(6) During cleaning harmful fluxes and residues are removed from the finished PCB. The required equipment is a cleaner. It can be online or offline.

(7) During inspection the soldering and assembly quality of finished PCB products is checkal. The required equipment for detection is magnifying glasses, microscopes, in circuit testers (ICTs), flying probe testers, automated optical inspections (AOIs), X-ray inspection systems, functional testers, etc. According to inspection requirements, its location in a production line is flexible.

(8) During rework the failed PCB product is reworked. Rework tools include a soldering iron, rework stations, etc. It can be in any appropriate location of the production line.

The four major pieces of equipment for the SMT process are screen printer, "pick and place" machine, AOI, and reflow oven.

Currently, the most popular assembly processes are all SMT and hybrid assembly, which has high integration density and uses both SMDs and through-hole devices. To balance the quality and cost, the reflow and wave soldering processes are the key factors.

7.2.3   Material and Cleaning Processes

Common materials for SMT are adhesive, solder paste, flux, cleaning agent, and anti-oxidation elements. The cleaning process is the basic process in SMT. Without proper cleaning, the flux residue will corrode the PCB. Common cleaning agents include CFC-113 (trifluoromethyl trichloroethane) and methyl chloroform. In addition, stabilizers, such as ethosome, acrylate, or epoxy type compounds, should be added during usage.

The basic requirements of cleaning agents are high degreasing ability, high dissolvability toward colophony and lipid, noncorrosive to metal, do not dissolve high molecular weight compounds, low surface tension, good wetting ability, can be removed from PCB in room temperature, low toxicity, nonexplosive, nonflammable, harmless to humans, stable, and does not react during the cleaning.

Common cleaning methods include soaking, ultrasonic cleaning, air cleaning, and spraying.

In recent years, the development of no-clean solder makes the no-cleaning process popular in SMT, and here are the reasons:

(1) The waste water produced from the cleaning process contaminates rivers and soil, and harms plants and animals.

(2) Solvents that contain chlorofluorocarbons compounds pollute the air.

(3) The residues from solvents are corrosive and lower product quality.

(4) No-cleaning reduces process time and saves money.

(5) The process eliminates the possiblity of damages in the cleaning steps.

(6) The electrical performance of residual flux has been improved and will not affect the product quality.

(7) The no-cleaning process has passed numerous international safety standards, and the no-cleaning flux residues are proven to be stable and noncorrosive.

7.2.4   Surface Mount Components

Surface mount components, abbreviated as SMC or SMD, can be of rectangular, cylindrical, cubic, or irregular shapes. SMD includes packaged semiconductor devices and bare chips. The soldering points or leads of SMD are in the same plane so they can be surface mounted on a PCB. SMD has undergone tremendous growth. Miniaturization is its main trend—it has moved from 1206, 0805, 0603 to 0402 or 0201 foot prints.

Surface-mount components have the following characteristics:

(1) They are small in size, lightweight, and can be mounted on both sides of the substrate for high-density package. For example, the size of a traditional transistor transceiver was reduced to less than 5 mm after adopting SMT.

(2) They have short leads or are leadless, reduced parasitic capacitance and inductance, and improved the high frequency performance.

(3) With their compact structure, SMT products have better vibration and shock resistance and higher reliability.

(4) The standardized shapes and footprints, the automatic assembly, the reflow process, the speed and high quality of production, the ease of adaptation for mass production and online inspection all contributed to lower overall cost of ownership.

Common surface mount components (SMCs) includes chip resistors, thick-film resistors, network resistors, film resistors wire wound resistors, thermistors, pressure sensitive resistors, tantalum chip capacitors, film capacitors, chip inductors, chip filters, chip oscillators, chip delay lines, chip switches, and relays. the SMD in the form of packaged semiconductor chips includes small outline transistor (SOT), small outline package (SOP), flat pack (FP), plastic leaded chip carrier (PLCC), quad flat pack (QFP), ceramic packaged device, leadless chip carrier (LCCC), leaded ceramic chip carrier (LDCC), etc. SMD chip assembly includes ball grid array (BGA), chip size packaging (CSP), etc.

7.2.5   SMT Design

The SMT design should consider various system requirements such as functionality, power consumption, frequency range, power input conditions, etc. Before going for SMT layout, the design should also consider components, substrate, and process selections. The basic rules for SMT design are as follows:

1. Circuit Partition Principles

(1) It has a block design according to function.
(2) It has separate analog and digital circuits.
(3) It has separate high-, middle-, and low-frequency circuit design, when necessary high-frequency part should be shielded.
(4) It has separate high power circuits from other circuits, for better heat dissipation.

2. SMT Substrate Design Principles

Substrate can be designed in "puzzle" fashion.
(1) The "puzzle" can be assembled by the similar or different circuit boards.
(2) The maximal size is determined by a "pick and place" machine and reflow oven parameters.
(3) A working area of 3–4 mm from the edge should be provided, and fiducial marks on the opposite corners should be designed.
(4) Each individual board is routed, which leaves the connecting bridge with the correct size and proper strength after V-grooving.

3. SMD Layout Principles

(1) The arrangement of SMD should follow its footprints; the same type devices should

line up in the same direction for ease of placement and inspection. This is also optimal for automatic pick and place. One carefully designed PCB is shown in Figure 7.6.

(2) Components, axes should be parallel or perpendicular to each other.

(3) The distribution of components should be even and give plenty board area for power devices.

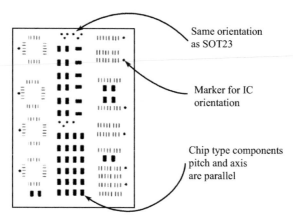

Figure 7.6   A SMD schematic layout

4. Pad Layout Design Principles

There are various pad patterns. The pad design will affect the reliability of the soldering joints, and the datasheet will show the correct pad pattern and parameters. Here are the general rules for pad layout.

(1) The pad pitch should equal the corresponding lead pitch.

(2) The pad width should equal the corresponding lead width plus or minus the K factor, which is determined by the component tolerance and placement accuracy.

(3) The pad length is determined by the lead's height and width and by the lead to pad soldering area. Pad length plays a greater role in solderability than the pad width.

7.2.6   SMT Testing

SMT testing techniques can be divided into noncontact testing and contact testing. The noncontact testing has evolved from visual inspection to automatic optical inspection (AOI) and automatic X-ray inspection (AXI). The contact testing can be further divided into online-testing and functional testing. The basic SMT testing methods include the following:

1. In-Circuit Tester (ICT)—The Most Common Electronic Testing Instrument

The traditional in-circuit tester uses the custom made bed-of-nail to contact with test points on the finished PCB and applies less than 1 volt voltage and less than 10 mA current to conduct isolated tests to determine various electrical parameters of onboard components, such as resistor, inductor, capacitor, diode, transistor, thyristor, field effect transistor, IC module, ASIC, etc., and failure modes such as missing component, misplacement, out of spec ification, soldering short, joint open, and related specific components and locations. The bed-of-nail in-circuit tester is fast and suitable for the single configuration mass production testing and costs less. However, with increasing PCB density and smaller SMT pitch, particularly the shorter R&D and production ramp up time coupled with the increasing variety PCB types, the bed-of-nail in-circuit test lags in the following aspects—long lead time to make bed-of-nail fixture, long trouble shooting time, expensive, and high-density SMT assembly can't be tested. The flying probe was developed based on bed-of-nail with probes replacing the pins. A typical flying probe station has four heads with eight probes on it in the x-y stage, and its minimal testing pitch is 0.2 mm. The probes touch the testing points according to the preprogrammed coordination and sequences and conduct the functional test of various components and its open and short conditions. Therefore, the flying probe station outperforms bed-of-nail in test accuracy, minimal pitch, and not requiring custom made fixtures, and testing programs can be generated by PCB CAD software. The disadvantage of the flying probe is the relatively slow test speed.

2. Functional Tester

ICT effectively searches various failure modes in the SMT assembly. However, it can't evaluate the performance of the whole PCB system. The functional tester can determine if the system fullfils its design specifications. It treats each target unit as one functional block, and by inputting various signals and testing outputs, it follows the design requirements to ensure normal operation. One simple functional test is to measure output signals after powering up the PCB assembly along with its supporting circuits. If the circuit assembly works normally, it passes the test. This method is simple and costs little. However, the failure mode cannot be detected automatically.

3. AOI

With increased packaging density, the contact electronic testing is facing increasing challenges, and the incorporation of AOI into online SMT testing is an important improvement in testing techniques. AOI is capable of inspecting not only soldering quality but also the quality of light pipe, solder paste, and placements. The AOI can replace most human operations and increase both product quality and production efficiency. The AOI camera automatically scans various portions of the PCB, compares the solder images with data from the library, determines PCB defects by image processing, and marks the defects on the screen for trouble shooting. The modern AOI system has adopted advanced computer vision, new lighting schemes, increased magnification, and complex algorithms, which enables fast and accurate error detection. AOI detects the following errors–missing components, wrong polarity of tantalum capacitor, misplacement, tilt, bending, folding, excess or insufficient solder, solder bridging, cold solder, etc. In addition to its ability to determine defects beyond human vision, AOI also gathers the product quality and defect mode data in all manufacturing processes for further analysis. However, AOI can't inspect electrical failures and invisible soldering points.

4. AXI

The AXI process is rather simple–after a PCB is fed into the machine, the X-ray is emitted by the X-ray tube above the PCB, goes through the board, and is detected by detectors underneath. Solder contains a large amount of lead, which absorbs X-ray. Therefore, the X-ray passing through the solder is much weaker than that of fiberglass, copper, or silica; thus the X-ray image for the solder is sharp, which makes it easy to find solder defects after simple imagine analysis. The AXI progressed from 2D to 3D inspection, the 2D version transmits X-ray inspection, which can generate a sharp image of the solder point on a single-sided PCB. However, for the double-sided SMT PCB, the solder joints on both sides will overlap and are difficult to separate. The 3D inspection uses tomography technology, which focuses the X-ray to the desired layer and projects the image to a rotating plane, where the image in the focal plane is sharp while images from other layers are blurry, thus enabling separate images of solder on each side. The 3D X-ray tomographic technology can not only inspect double-sided PCB, it is also capable of observing invisible solder, such as those in BGA, and give "sliced" images of BGA–solder's top, middle, and bottom. In addition, it can also inspect through-hole soldering, such as the filling of a hole, thus greatly improving the soldering quality.

7.3　Packaging of Flat Panel Display Modules

7.3.1　Introduction of Liquid Crystal Display

Series products of displays, one of the most commonly used human–computer interfaces in modern information society, are very important for civil, commercial, and military use. Products related to people's daily life are flat televisions, laptops, cell phones, personal digital assistants (PDAs), etc. Information technology would not have developed so fast without display monitors. Display technology, containing electrical, communication, and information processing technologies, is considered another opportunity for the electronic industry after microelectronics and computers in the 20th century.

With the rapid development of science and technologies, display technology has been a revolution. Especially since the 1990s, the sharp increase in technologies and market demand brought on the rapid development of flat panel display (FPD), with liquid crystal display (LCD), among others.

According to the time of technology development, LCD has been through three stages. From the 1970s to the early 1980s, twisted nematic LCD (TN-LCD) was popular, which had a simple structure and fabrication process. But it had very poor display capacity and was mainly used in watches, calculators, digital displays, and similar simple electrical products. Advanced TN-LCD was used in meters, cameras, telephones, mobile meters, sound boxes, and so on. In the mid 1980s, super twisted nematic (STN-LCD) LCD was developed. With its high performance, large display capacity, and low cost, it was extensively used in automated office products and communication consuming products, such as cell phones, PDAs, GPSs, laptops, pagers, digital dictionaries, electronic diaries, learning machines, and so on. In the 1990s a new thin-film transistor LCD (TFT-LCD) was developed. Through sputtering or the chemical deposition process, various films are fabricated on glass or plastic substrate to make a large-scale integrated circuit in TFT. The cost decreased sharply using non–single crystal substrate, and conventional large-scale integrated circuits have been developed as pioneers of large area, multiple functions, and low cost. It is more difficult to fabricate controllers to switch on/off pixels (LCD or LED) on glass or plastic substrate with a large area in TFT than to fabricate large-scale IC on silicon. Requirements for the fabrication environment (class 100 level), for material purity, and for manufacturing instruments and techniques are superior to those for large-scale ICs, since they are the most advanced technologies of modern mass production. In applications, the bottleneck of TFT-LCD is to overcome the disadvantage of long response times for STN-LCD, and it also presents high-quality color displays and flexible display sizes. It is widely used in digital cameras, camcorders, televisions PCs, and especially in laptop computers.

Since the 21st century began, STN-LCD has become more colorful, high precision, and thin. TFT-LCD takes the lead in the portable electronics products market. It is predicted that with lower cost, TFT-LCD will dominate the market in a few years.

7.3.2　Packaging of Display Modules

With the development of IC and interconnect technologies, liquid crystal modules (LCMs), consisting of LCD, driving IC, and controller IC, are not only convenient for customers to use directly in products, they also increase the added value of LCD, as an extension of LCD products. Display modules were developed through several stages: In the early 1980s, SMT was developed and mainly used in various instruments and meters; since the mid 1980s, chip-on-board (COB) display modules emerged and were mainly used in various portable electrical products; in the early 1990s, chip-on-glass (COG) display modules, tape automatic bonding (TAB) display modules, and chip-on-foil (COF) display modules were gradually developed in Japan and the US, directing the industry.

The demand of the display module market increased rapidly with the development of LCD, mainly in communication display modules such as mobile phone applications. A display module is a module that uses a display screen as the main body and has complete display functions. In terms of clamshell cell phones, in addition to main display and sub-display modules, sound and video module and other electronic parts on the cover are also included. Therefore, a typical display module includes screen, driver chip and its associated components, backlight, sound (horn, microphone, shaker), video (camera, image processor), etc. The requirements for a display module are given by the customers according to their requirements for structure, function, and cost. Figure 7.7 is a picture of a typical assembled display module. This module connects the main board of a cell phone through a connector assembled on flexible substrate. The core part of the display screen and driver are shown in Figure 7.8.

Figure 7.7   A typical display module

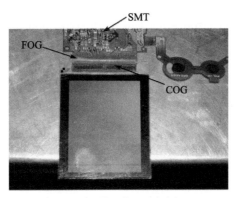

Figure 7.8. Display and driver

From Figure 7.8, it can be seen that the driver chip is mounted on the glass as part of LCD using chip on glass (COG) technology. The passive components are mounted on flexible substrate using surface mounting technology. The flexible substrate is connected to the glass through the foil-on-glass (FOG) technique using the thermal compression process. Therefore, the main techniques to be introduced here include COG, COF, and FOG.

7.3.3   Chip-on-glass (COG) Packaging

COG is a typical flip-chip packaging process. In 1983, Citizen Co. in Japan first announced it was using COG packaging in the palmtop television LCD.[7] After that, many manufacturers have developed various COG packaging methods. COG has the advantages of high density, low cost, high efficiency, and small size. Especially in recent years, with the developments in the density and capacity of flat panel displays, COG packaging has become dominant in display driver chip packaging.

Like other flip-chips, COG needs bumps to be fabricated on the active surface of chips as interconnection pins. The bumps used for COG packaging are mostly "upright wall" Au bumps fabricated with the photolithography technique. The electrodes on glass substrate usually use indium tin oxide (ITO) transparent conductive material. Because of the restriction of size and layout of flat panel displays, COG basically uses slender chips, the bumps of which align along two sides in parallel or interlace and mostly have rectangular cross-sections. Usually the distance between the centers of adjacent bumps is called pitch, and the gap between adjacent bumps is called gap, which are two main parameters for the density of COG packaging, as shown in Figure 7.9.

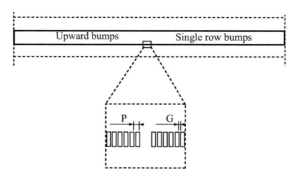

Figure 7.9   A schematic structure showing the pitch and gap

Flip-chips are divided into two main types according to the method of interconnecting between bumps and electrodes on the substrate, namely, metallurgy contact and mechanical bumps and corresponding electrodes reacting with solder joints. Under these conditions, bumps of chips are usually fabricated in solder material. In the soldering process, after an IC is mounted on a substrate, the reflow process will connect the IC to the substrate through the formation of solder joints. After reflow, underfill is usually applied to seal and protect the interconnections and the chip. Mechanical contact, by definition, is electrical conduction between the bumps and corresponding electrodes through mechanical contact where the connections are maintained by additional adhesive. Usually, mechanical contact may be direct contact between bumps and electrodes or through other conductive materials, such as anisotropic conductive film (ACF). Owing to the limitations of temperature, cleanness, efficiency, and cost of fabricating flat panel displays, with the exception of some reports on the development of COG eutectic bonding by some researchers and institutes, all COG processes are mechanical contact film (ACF), isotropic conductive adhesive (ICA), or nonconductive film (NCF).

1. ACF Bonding

ACF is an anisotropic conductive thin adhesive film, which completes interconnection and adhesive curing at the same time in the packaging process. ACF is currently the most widely used packaging material in COG. ACF is fabricated by disseminating uniformly 0.5%–5% conductive particles in a polymer matrix. In the COG process, ACF is first placed onto the glass substrate, covering all of the ITO electrodes. A driver IC is then mounted on the glass substrate through alignment, prebonding under certain temperature and pressure conditions. At last the main bonding is completed under higher temperature and pressure conditions, which needs typically 5–10 s to fully cure the adhesive. Therefore, the main bonding is the critical step of COG packaging in terms of efficiency. Dispersed conductive particles in ACF usually consist of an elastic polymer core coated by Ni/Au metal layers. In the thermocompression process, some conductive particles are captured by bumps and corresponding ITO electrodes and deformed to form the conductive path between bumps and electrodes. In the other direction, because of low distribution, conductive particles cannot form conductive paths, so the anisotropic interconnection is established, as shown in Figure 7.10. A typical conductive particle is resin ball plated with Ni-Au, as shown in Figure 7.11. The balance of stress is the main consideration.[8] In an ACF bonded component, the compression stress provided by the adhesive and the elastic stress of the conductive particle are balanced, which is good for reliability. The polymer in ACF is commonly thermoset, which cures by heat in the bonding process. More than 80% curing is typically required for good mechanical and electrical properties of COG packaging. The main function of the adhesive is to attach the die and seal the COG. In addition, since the glass substrate is transparent, some ACF in COG is photodefinable material.[9] This kind of ACF cures in ultraviolet light during packaging. The COG process is thus performed at room temperature, which is the most outstanding advantage. This type of material is particularly suitable to the low-temperature COG process.

To sum up, the advantages of ACF are as follows:

(1) Fine pitch flip-chip interconnection.

(2) No underfilling.

(3) Low process temperature.

(4) Simple, flexible, and low cost.

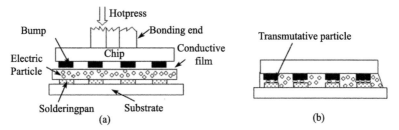

Figure 7.10   A schematic view of an anisotropic interconnection

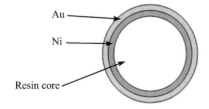

Figure 7.11   Conductive particles in ACF

2. ICA Bonding

Isotropic conductive adhesive is a packaging material that has the same conductivity in all directions, a typical composition of which is polymer with Ag filler.[10] The ICA bonding process is similar to the soldering process. First the ICA is printed on the electrode pads on the substrate. Then the chip is aligned and mounted onto the substrate, the polymer in the ICA cures through thermal compression to bond the chip onto the substrate. At the same time, conductive filler in the ICA forms a conducting path between the chip and substrate. At last the underfill is applied between the chip and substrate for sealing and protection. Unlike soldering, ICA packaging doesn't need to reflow.

ICA is one of the first packaging materials developed in the COG process. Citizen Co. in Japan first used ICA in the COG packaging of palmtop liquid crystal displays.[10] In the process, bumps of IC chips were Au plated Cu mushroom bumps, with a pitch of 200 μm. In addition, they used ICA in the production of a small-sized LCDs with a bump pitch of 150 μm.[11] The ICA was transferred and attached to the bumps, instead of printing on the ITO electrodes of the glass substrate. The ICA-COG process developed by Panasonic Co. in Japan used stud bump technology.[12] The fabrication of stud bumps is similar to the wire bonding process. Metal wire fuses to a ball on the chip pad and is drawn to break on top by bonding force to form nail head bumps. Then all bumps are pressed to the same height. Compared to plating bumps, this kind of fabrication of bumps is simple and cheap. The diameter of metal wire used in Panasonic Co. fabrication was 20 μm, so that the size of the square pad could be reduced to less than 70 μm, which is the pitch of bumps (Fig. 7.12).

Compared to ACF packaging, the ICA bonding process is relatively more complicated with lower bonding strength. Especially because the contact area between ICA and the chip and ICA and the substrate is small, many chips are detached before underfilling. In addition, high mechanical residual stress remains inside devices after underfilling, which will cause product failure in reliability tests. Therefore, up to now ICA has been limited in application to packaging small size and large bump pitch products.[13]

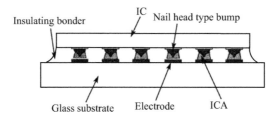

Figure 7.12   A schematic of stud bump bonding

3. NCF Bonding

The function of NCF (Nonconductive film) is similar to ACF, which is to establish electrical conducting paths between different components for interconnection.[10] Unlike ACF, which introduces new conductive particles as a conducting medium, in NCF, all conductive particles are removed and the electrical connection is established by direct contact between chip bumps and ITO on substrate through compression. After the thermal compression process, thermoset nonconductive film can provide mechanical fixation for the direct contact to ensure the stability of the electrical connection. Because NCF is used to ensure the connection of an IC and substrate both mechanically and electrically, the curing condition is very important for NCF.

In an NCF-bonded device, stress is an important factor for electrical and mechanical interconnection reliability. Thermal stress, binding force, and internal shrinkage provide the bonding force of the assembly and the compression force between bumps and the ITO. Because of the low elastic deformation of glass, the elastic deformation of the bumps is extraordinarily important for releasing the internal stress of the device. Experimental results showed that the bumps with better elastic properties performed much better than those bumps with low elasticity in reliability tests.[14] One of the key requirements of NCF bonding is that every bump must be compressive to ensure the electrical conductivity of the interface after bonding and in service conditions.

NCF bonding avoids possible electrical shortages between bumps, as in the case of ACF bonding. Therefore, it could be an interconnection technique for ultra–fine pitch IC bonding. However, the current bumping technology cannot ensure the planarity of the bumps, which is the requirement for bumps to be under compression after bonding and during service conditions. Therefore, the NCF technique has not been used in mass production.

7.3.4 Chip-on-foil (COF) Packaging

COF is similar to COG in process but different in substrate material.

Based on the lamination technique between substrate and copper foil, a flexible circuit board is divided into two types: a flexible board with adhesive and an adhesiveless flexible board. The adhesiveless flexible board is much more expensive than that with adhesive, but its flexibility, bonding force between copper foil and substrate, and flatness of pads are better than that with adhesive. Therefore, it is commonly used to fulfill stringent requirements. COF uses adhesiveless flexible substrate. This is because the high bonding temperature will soften the adhesive, which may result in the copper tracks on top of it being unstable. This greatly increases the possibility of bonding misalignment.

There are two methods to fabricate adhesiveless flexible substrate:

1. Coat Polyimide on Copper Foil

Screen print polyimide solution on copper foil and vaporize the solvent at high temperature to cure polyimide; at the same time, the cured polyimide layer is bonded to the copper foil. This method needs a certain thickness of copper.

2. Plate Copper on Polyimide Layer

First a layer of metal film is deposited on polyimide film through sputtering or evaporation. Cu is then plated to an expected thickness. Because the space between copper wires is related to the thickness of copper foil (less space for thinner foil), compared to the first method, copper foil has no thickness restriction.

Like COG technology, ACF is extensively used in COF. The process principles and procedures are similar to COG technology, too. The main differences are as follows:

(1) The adhesive must match with the substrate for bonding (glass or flexible substrate). The selection of the thickness of ACF should consider the copper wire thickness on flexible substrate.

(2) Because of the high coefficient of thermal expansion and the high bonding temperature of COF, compensation for the thermal expansion mismatch between IC and substrate should be considered in the design of the flex circuit, in particular, when the IC size is large.

Like COG, when the gap between bumps on the IC becomes small, the possibility of a short circuit between bumps in ACF bonding increases. NCF technology and the thermocompression process can avoid or relieve this problem. These technologies have good compatibility with ACF technology on the equipment and can be used for fine-pitch interconnections.

NCF bonding works on the same principles in COF and COG. The flexible substrate can be compensated to a certain degree with the height difference between bumps. Thus NCF is an excellent technology for COF.

The thermocompression process uses the diffusion bonding between Au bumps of die and the plated Ni/Au layer of printed wires on the substrate or eutectic bonding between Au bumps and the plated Sn layer of printed wires on the substrate to establish mechanical and electrical interconnections.[15,16] Underfill is applied to protect and strengthen the connections. The bonding temperature of the Au-Sn alloy is approximately 250°C, and the bonding temperature of Au-Au thermocompression is approximately 330°C. The process of Au-Au thermo compression is shown in Figure 7.13.

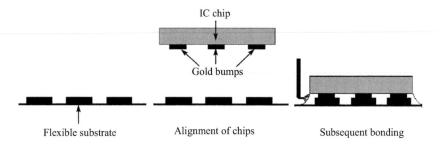

Figure 7.13   The process of Au-Au thermocompression

There are some issues of NCF and thermocompression bonding for COF. A high bonding force is needed on the IC chip to ensure that when the force is removed, the cured NCF can put the bumps in their places, where the mechanical and electrical connections are already established. The high force may damage the Al pad underneath each bump at the moment of bonding due to stress concentration, which may affect the electrical function and the reliability of the service condition. The same problem exists in thermocompression bonding as in NCF bonding. That is, a large bonding force is applied to the IC chip to ensure the formation of diffusion or eutectic bonding. Therefore, the design of the copper track's size on the flexible substrate corresponding to bumps on the IC has to be considered together with the optimization of the bonding process parameters, especially the bonding force.

7.3.5   Foil-on-glass Thermal Compression Bonding

Bonding of foil-on-glass (FOG) is also done during the thermocompression process using ACF. Ni/Au is plated on copper tracks at the bonding location of the flexible substrate. The thickness of Ni for thermocompression bonding is typically 1–2 μm. The thickness of Au is 0.05–0.1 μm. To increase bending resistance, Au may be plated directly on copper tracks.

When bonding flexible circuit boards with copper foil-on-glass with ITO, the thickness of ITO is neglected compared to the thickness of the copper foil. Usually, because the pitch between copper tracks at the flexible substrate-glass bonding location is small, the alignment between the flexible circuit board and glass is performed manually or automatically with the help of a vision system. Generally, when the gap of the tracks is less than 0.1 mm, a heat seal machine must be used with automatic vision alignment.

The curing process of ACF material is completed by a heat seal machine with a "fixed heating bar." Because the thickness of copper foil is commonly between 8 μm and 36 μm, the thickness of ACF is chosen according to the thickness of copper foil. Generally it can be calculated with formula (7.1):

$$t_0 = \frac{S_1 + S_2}{2 \times P} \times T + t_1 + \alpha, \tag{7.1}$$

where t_0 is the thickness of ACF (μm); t_1 is the thickness of ACF between ITO and copper foil after packaging; T is the height of the copper foil; P is the pitch of the Cu tracks; S_1 is the space between the top of the Cu tracks; S_2 is the space between the bottom of the Cu tracks (Figure 7.14); and α is the correction parameter $(0.26 \times T)$.

When bonding a flexible substrate on glass, the number of pin counts can be hundreds and the total bonding length can be more than tens of millimeters. The mismatch in thermal expansions between glass and flexible substrate when bonding at high temperature may induce the misalignment between ITO and corresponding copper pins. To avoid this phenomenon, compensation in thermal expansion needs to be considered in the design of flexible substrate I/O pin positions. The amount of compensation is determined by the coefficients of thermal expansion of flexible substrate and glass as well as the bonding temperature.

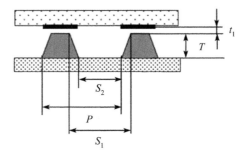

Figure 7.14   Schematic structure of FOG

7.4   Optoelectronic Packaging

Optoelectronic packaging design consists of optical, electrical, thermal, and mechanical aspects of the system level. Among them, thermal consideration is critical. This is partially owing to the optoelectronic device's sensitivity to temperature. For example, to make optical components work properly at specific wavelength, temperature control is normally needed, and in some cases, a proper cooling device is also added. Future optoelectronic devices will have higher density and switching speed, which all make thermal management indispensable.[17−24]

There are various optoelectronic devices and packaging styles. In the following sections, we start with the common laser diode and LED packaging and then discuss some special optoelectronic packaging processes.

7.4.1   Laser Diode Packaging

Laser diode packages can be divided into three categories, namely, transistor outline thermoelectric cooler (TO-CAN), DIP, and butterfly package.

The TO-CAN package is popular among low transmission rate devices, which is in the range of 100 Mbps to 200 Mbps where the packaging is simple and without thermoelectric cooler (TEC). In this case, no temperature control is needed. Because of its low cost, simple structure, and suitability for volume production, TO-CAN is now widely used in data communications, LAN, optical time domain reflectometer (OTDR), and CD and DVD players.

Figure 7.15 is the internal structure of a TO-CAN package. The package combines the laser diodes, power ground, and photodetector's output ground into one pin.

The DIP package can transmit up to 1 Gbps. In addition to its high speed, it has a bigger footprint, a larger heat dissipation area, and can fit in a built-in TEC and a sensor for higher power output. With hermetic capability, DIP can also be used in harsh environments.

Because of its cost, the DIP package is normally used in communication, cable TV, LAN, OTDR, etc. Figure 7.16 is a typical example of DIP's internal structure.

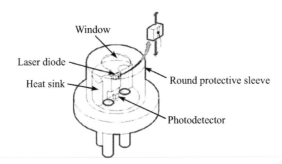

Figure 7.15   Internal structure of a laser LED packaging

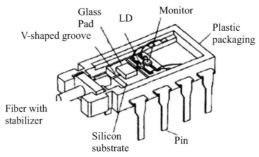

Figure 7.16   The inner structure of DIP

The butterfly package is a type of DIP in principle, as shown in Figure 7.17, and shares the advantages of DIP. However, owing to shorter pins, its speed can reach up to 2.5–10 Gbps.

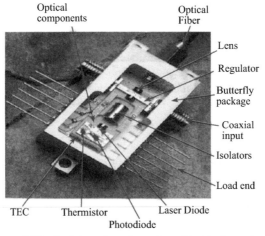

Figure 7.17   An internal configuration of butterfly package

The following steps are for butterfly packaging. First, the heat sink is soldered to the metal base; then the laser diode chip is brazed on the heat sink; then the fiber core is aligned to the output of the laser diode till the maximum light coupling is obtained; and finally the regulator is brazed to the metal base.

7.4.2   LED Packaging Technology

LED is a device that can convert the electrical energy directly to either visible or invisible light. After over 30 years of development, we can produce the entire visible spectrum

LED with high brightness and high quality. LED packaging technologies were developed from discrete device packaging technologies and adapted for the optoelectronic applications. Normally, a discrete device die is sealed in the package in order to protect the die and make electrical connections. However, LED packaging is used to complete the electrical connection, protect the die, and get the optical signal out. Therefore, LED packaging cannot use IC packaging technologies directly. The materials for LED packaging include metal alloys, glass, polymer, and binding materials. Physical properties such as the coefficient of thermal expansion, thermal conductivity, glass transition temperature, creep, stress-propagation, and curing properties will all affect the stability and reliability of the packaged module. In addition, the packaging format also affects the performance. Incorrect choice of material or packaging style will compromise the long-term reliability.

As shown in the Figure 7.18, the traditional LED package includes the chip and its surrounding epoxy resins. When the current flows into the semiconductor's P-N junction, electrons and holes recombine, and light emits. However, the light from the junction needs to transmit first through the internal lattice of the LED chip, then through external packaging materials. The package is not only for protection, the reflective cup of the lead frame also acts as a reflecting and focusing mirror. The bullet type epoxy resin functions as a focusing lens. This type of package is a typical LED point light source, which is suitable for indication lamp applications.

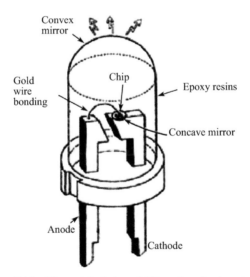

Figure 7.18   Through-hole type LED and packaging structure

In production, LED packages have gradually evolved to a variety of forms, such as Φ3 mm, Φ5 mm, Φ8 mm, and the square-head. Owing to the refractive index mismatch between the surrounding packaging material and the LED chip, the light produced from the LED junction will be partially reflected and absorbed, which greatly reduces the LED luminous efficiency. Because the body diameter is 3 mm or 5 mm, those LEDs were also named T1 and T3/4. Although T1 and T3/4 are low-cost packages and have excellent light emission patterns, they are still too big for some applications. Developing new material to improve LED's luminous efficiency is the focus of contemporary research, especially in the general lighting areas of which often require tens or hundreds lumens of light. One T1 or T3/4 LED device can only give one or two lumens. Though it is possible to bundle a large number of LEDs to achieve higher light output, the heat generated is hard to dissipate and it reduces reliability. With requirements for a small footprint and higher power output, the traditional

LED packages are no longer suitable for applications with 100 mA and above. Therefore, problems with heat dissipation and cost have slowed the proliferation of LED products.

In early 1980s, to reduce manufacturing costs, the LED industry began to use automated assembly technology for surface mount devices (SMDs), and the trend continued into 1990s. The size and weight of a SMD LED is only 1/10 of the traditional leaded LED. Therefore, SMD LEDs are widely used in a variety of mobile devices.

In 1983, Citizen manufactured the first commercial SMD LED. The initial product was a low-power device for indicator lights and keypad lighting of mobile phones. Later it evolved to more powerful SMD LEDs for automotive dashboard lighting, brake lights, and special and general lighting equipment. After 20 years of development, SMD LEDs have become standard product offerings, with the market share just below leaded LEDs.

SMD LEDs are normally 1 mm or 0.8 mm thick. To meet the demands of cell phones, PDAs, and other portable electronic devices, researchers have developed the chip-type LED, with thicknesses of 0.6 mm, 0.5 mm, and 0.4 mm.

Die attachment, wire bonding, and epoxy sealing are the three critical steps in LED packaging. An automatic die attaching machine is normally used for die attachment. First, conductive paste is dispensed onto the substrate; then the die is attached onto the conductive paste; then the electrodes of the LED die are wire bonded to the lead frame. Since the shape of a LED lens relies on the molding shape, a variety of molds are made for various lens shapes. The casting method is popular for leaded LED mass production, since the lens follows the external molding shape precisely and molding can be changed easily. On the other hand, because SMD LEDs are normally of standard size, transfer molding is more popular. In transfer molding, the LED dies with wire bonding are inserted into the mold in reverse; resin is injected, heat cured, and hardened; then the finished LED is removed.

Multichip packaging is an effective method for power LED packaging. For example, the power LED, developed by UOE Inc. in America in 2001, uses hexagonal aluminum as the substrate for efficient packing.

The hexagonal aluminum is 1.25 inches in diameter. The luminescence is at the central location, with a diameter of around 0.375 inch. It can accommodate 40 chips. The aluminum is used as a heat sink while electrically insulated. The anode and cathode of the chips are connected to the corresponding pads on the substrate by wire bonding. The number of LED dies on each substrate is determined by the output power requirement. The combination includes ultra high brightness AlGaInN or AlGaInP chips, with monochrome, multicolor, mixed color, etc. By using the high refractive index material and optimized lens shape through optical design, high light extraction efficiency can be achieved while the chip and bonding wires are protected.

7.4.3   Optoelectronic Coupling and Fixation

The key technologies of optoelectronic packaging are optical coupling and the placement, especially the optical coupling and placement among different types of devices. The quality of the coupling and placement determines whether the device has a high optical coupling and high reliability, which thus determines the ultimate performance of the device. As each optical device has six degrees of freedom (three translational and three rotational degrees of freedom), it is difficult to achieve optimum optical coupling and placement during packaging.

1. Optical Coupling and Alignment

In the process of connecting fiberoptic to optoelectronic devices, there are many unpredictable factors, such as fiber composition, end-plane flatness, standardization of components, etc. The unpredictable factors are also the tolerances in electromechanical and optical properties. In order to achieve a minimal optical attenuation, it has to go through optical alignment.

Optical alignment can be classified into active and passive alignment. Active alignment uses external instruments (such as the multiaxis servo) to find the best light alignment, and then fix the device permanently. The traditional method of active alignment is to clamp down the optical components and then make translational movements in the XYZ direction and the rotation between the two devices. This process relies on a light searching algorithm in the alignment, and the quality is determined by the algorithm. Generally speaking, active alignment is simpler and has lower technological and equipment requirements—it only requires changing the holding fixture for different devices—and the search algorithm can be reused. However, the drawback is its relatively longer alignment time. Table 7.1 lists the technical parameters and attributes of active and passive alignment.

Table 7.1 Comparison between active and passive alignment

	Active alignment	Passive alignment
Alignment accuracy	Less than a micron	More than a micron
Alignment method	Uses tools to achieve the best alignment	Uses semiconductor etching technology to create a V-shaped groove to store fiber
Alignment time	Long	Short
Light insertion loss	Low	High
Yield determination	Chip selection and electrical testing	Batch inspection
Repeatability	OK	Good
Lifetime	OK	Good

Passive alignment is accomplished when both emitting and receiving units are on. We can get physical alignment through lithography and the conventional machining process. By simple insertion, the optical axis can be aligned correctly without needing adjustment and can be produced in volume. The V-shaped groove (V-groove) has been developed with an alignment accuracy of 0.5 μm, as shown in Figure 7.19. With the help of markings, the alignment accuracy can reach 0.3 μm. Strategies include embossing chimerical alignment, guided insertion alignment, and other structure matching alignments. The biggest advantage of passive alignment is short coupling adjustment time. The disadvantage is the need to redesign grooves whenever a new process or different component gauge is needed for a new application. Making the V-groove requires high accuracy and repeatability; otherwise the yield will be low. In addition, in passive alignment, the insertion loss can only be checked after component assembly. Passive alignment is still the focus of current research

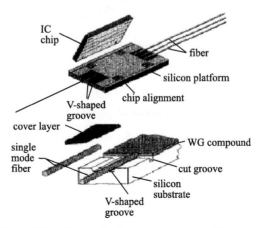

Figure 7.19 The packaging structure using V-groove

in many companies. With readily available semiconductor etching, passive alignment is the preferred method in mass production.

To prepare for passive alignment, a V-shaped groove needs to be etched on substrate for the placement of fiber, and fiducial marks need to be made on substrate for alignment of the device. Those steps can be accomplished by photolithography. In the assembly process, a device is placed on the substrate aligned through the fiducial marks. A light pressure is applied while the Au-Sn solder is heated to fix the device in the right position. Through accurate matching between substrates and optical fiber, self-alignment can be obtained. As explained above, passive alignment relies upon precise alignment among components, substrates, their markings, and various micromachined mechanical fixtures. In general, the passive alignment process can easily be automated. The two major challenges are micromachining and assembly accuracies.

In the optical communication component market, active and passive alignment has its own niche. There is a new method that combines active and passive alignment—in addition to passive alignment, it has added precision driving by electrostatic, thermaldrive, etc. After the fiber is in the V-groove, the precise position is achieved by a microactuator through minimizing insertion loss. This method has achieved an accuracy similar to active alignment. The current trend is to put more optoelectronic components into one package. Since the microactuator is readily producible by semiconductor processing technology, the microactuator is likely to be manufactured along with optoelectronic device fabrication, which is the future trend for high-volume and low-cost production. Figure 7.20 shows the optical fiber groove in a 2 × 2 optical switch for passive alignment, which was developed by National Key Laboratory of MicroNano Fabrication Technology at Beijing University. Within the optical groove, there is also a snapping structure for the optical fiber attachment, thus further improving the reliability of the passive alignment.

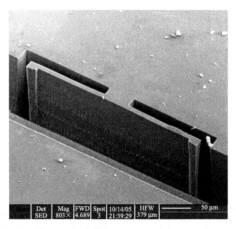

Figure 7.20   The optical fiber groove with snapping structure

2. Chip Fixation and Packaging after Alignment

The receptacles or pigtail modules are made for ease of usage and coupling stability. In the light source attachment, the key point is that the coupling between light sources is stable against time and environmental changes. The common ways to fix the light source are epoxy gluing, tin welding, mechanical attachment, electrical casting, laser welding, etc. In general, the transverse displacement tolerance is small in the single-mode fiber connection, where the laser welding is used, while the gluing method is normally applied in multimode fiber, where the transverse displacement tolerance is larger.

Questions

(1) What are the fundamental fabrication processes in SMT technology? What are the requirements for devices to be used in SMT?

(2) What are the special characteristics of optoelectronics packaging and corresponding packaging processes?

(3) What is special for glass packaging? How are they used in display module fabrications?

References

[1] Zhaohua Wu and Dejian Zhou. Basics of Surface Mount Technology. Beijing: National Defense Industry Press, 2003.

[2] Liding Zhang. Advanced Electronic Manufacturing Technology. Beijing: Publishing House of Elecronics Industry, 1993.

[3] Dalin Xu. Surface Mount Technology (SMT), its Trends and Future. Chinese Journal of Electronic Devices, 2(1999): 107–108.

[4] Shuzhen Li, E. Zhou, and Peng Guo. Surface Mounting Technology and its application. Space Electronic Technology, 1(1999): 64–66.

[5] Feng Lu. Development And Countermeasures of Our Country SMT Equipment. Electronic Process Technology, 3(1999): 39–41.

[6] Ynxiang Kuang and Songchun Zhu. Advanced Micro-electronics Packaging Technlogy in SMT. Electronics Process Technology, 4(2004): 40–44.

[7] Kristiansen, H. and Liu, J. Overview of conductive adhesive interconnection technologies for LCDs. IEEE Transactions on Components, Packaging and Manufacture Technology, 21.2(1998): 208–214.

[8] Myung-Jin Yim, Kyung-Wook Paik, Tae-Sung Kim et al. Anisotropic conductive film (ACF) interconnection for display packaging applications. Electronic Components and Technology Conference, 1995: 1036–1041.

[9] Chang Hoon Lee and K.I. Loh Fine pitch COG interconnections using anisotropically conductive adhesives. In: Proceedings of the 1995 Electronic Components and Technology Conference, 1995: 121–125.

[10] Johan Liu. Conductive Adhesives for Electronics Packaging. Port Erin, UK: Electrochemical Publications Ltd., 1999.

[11] Miyajima A. Small liquid crystal display device for projection. Large-screen Projection Displays II, 1255.46(1990): 46–51.

[12] Y. Bessho, Y. Horio, T. Tsuda, et al. Chip-on-Glass mounting technology of LSIs for LCD module. In: Proceedings of the International Microelectronics Conference, (1990): 183–189.

[13] J. Vanfleteren, B. Vandecasteele, T. Podprocky, et al. Low temperature flip-chip process using ICA and NCA (isotropically and non-conductive adhesive) for flexible displays application. Electronics Packaging Technology Conference, (2002): 139–143.

[14] Shyh-Ming Chang, Jwo-Huei Jon, Hua-Shu Wu, et al. Stress analysis of "Micro-Bump Bonding" structure for "Chip on glass" packaging. In: Proceedings of the 5th Asian Symposium on Information Display, (1999): 79–83.

[15] Kim Y.-G.G., Kumar Pavuluri J., White J.R., et al. Thermocompression bonding effects on bump-pad adhesion. IEEE Transaction on Components, Packaging and Manufacturing Technology-Part B, 18.1 (1995): 192–200.

[16] S.K. Kang. Gold-to-aluminum bonding for TAB applications. IEEE Transactions on Components, Hybrids, and Manufacturing Technology, 15.6 (1992): 998–1004.

[17] S. Haque, D. Steigerwald, S. Rudaz, et al. Packaging challenges of high-power leads for solid state light. IMAPS'3, Boston, MA, 2003: 100–105.

[18] Li Zhang, V. Arora, L. Nguyen, et al. Numerical and experimental analysis of large passivation opening for solder joint reliability improvement of micro SMD packages. Microelectronics Reliability, 44 (2004): 533–541.

[19] D. Madhav. Electrochemical processing technologies in chip fabrication: challenges and opportunities. Electrochemical Act, 48 (2003): 2975–2985.

[20] M. Arik, J. Petroski, and S. Weaver. Thermal challenges in the future generation solid state lighting applications: light emitting diodes. ASME/IEEE International Packaging Technical Conference, (2001): 113–120.

[21] Ganlin Zhou. The Research of the Alignment Algorithm for TO-CAN Package Laser Wielder Machine. Taiwan: National Cheng Kung University, 2005.

[22] Ruijun Zhang. Coupled alignment technology to package photoelectric device. Photon Technology, 2 (2003): 79–83.

[23] Wood-Hi Cheng, Maw-Tyan Sheen, Gow-Ling Wang, et al. Fiber alignment shift formation mechanisms of fiber-solder-ferrule joints in laser module packaging. Journal of Lightwave Technology, 19.8 (2001): 1177–1184.

[24] Kuang, J.H., Sheen, M.T. Chen, J.M., et al. Reduction of fiber alignment shift in add/drop filter module packaging. In: The 4th Pacific Rim Conference on Lasers and Electro-Optics, 2 (2001): 236–237.

CHAPTER 8

System-level Packaging Technology

8.1 Overview[1−3]

In recent years, packaging technologies have been seen from a systematic point of view. In a broad sense, system-level packaging, from a systematic optimization stance, tends to put chips and modules of the same or different types together, either to get prepared for the next level of integration or to directly implement a finished product. In a narrow sense, a microsystem may comprise chips of different types: digital circuitry, analog circuitry, I/O interfaces, memories, DSPs, CPU, and MEMS. In that sense, a pure memory can only be considered a memory rather than a system, no matter how high the degree of integration level is; moreover, the simple integration of multiple chips of the same categories, for the implementation of a new function, may not be thought of as a system. The concept of system involves the convergence of multiple technologies. Thus, the realization of systematic integration usually involves the use of different materials and the adoption of various architectures, and then demands the concerted collaboration of process, circuitry, and system engineers.

There are two mainstream technologies at present: one is system on chip (SOC); the other is system in package (SIP). SOC, the monolithic systematic integration, is considered the technology capable of maximizing the degree of integration. It evolved from the large-scale system chips, which combined CPU, DSP, memory unit, and accessory circuitry into the same chip.

In addition to the placement of formerly discrete components on a single chip, a system-level tradeoff-the control of internal signal transmission, allocation of memory components and the unification of a flow with heterogeneous natures-is also demanded. SIP is defined as systematic integration based on packaging techniques, which usually denotes the vertical stacking of multiple chips of a heterogeneous or homogeneous nature, assembled into the same packaging housing, with interconnection between the chips implemented during the same process.

SOC and SIP, as well as the radio frequency (RF) system packaging technology, will provide effective methods of enhancing the microsystem function and of further increasing integration, in the same way that MEMS and module assembly/optoelectronic packaging discussed in Chapters 6 and 7, respectively, did so. Whether it is carried out on the chip level or the package level, these cutting-edge technologies represent advancements in the microsystems domain and can facilitate the convergence of portable and desktop devices. At the same time, these packaging technologies also foreshadow a closer correlation between packaging and front-end wafer processing, signifying the codevelopment of both.

8.2 System on Chip Technology[3,4]

8.2.1 Overview

System on chip, usually referred to as SOC or SoC, is a concept that emerged in the 1990s, and whose definition has continuously been enriched with time and technical advancements. With 65 nm 12 inch wafer manufacturing foundries available, hundreds of million transistors have been integrated in one chip. Currently SOC can be defined as the monolithic integration of a whole set of systems, including basic circuit units, such as one or more processors, memories, analog circuitry modules, mixed analog/digital circuit modules, programmable logic units, etc. A schematic of the concept of SOC integration is shown in Figure 8.1. If pertinent design and process issues can be solved, SOC techniques can provide the availability of system products with the highest integration and weight. Meanwhile, packaging of this chip or microsystem only requires the provision of conventional packaging functions such as signal transmission, power supply, and cooling. Thus the implementation of these kinds of SOC products has been a target pursued by various equipment system and semiconductor vendors.

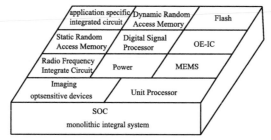

Figure 8.1   Schematic of the concept of SOC integration

Technologies for SOC include: need to define intellectual property, programmable system chip, development and application of core chips for IT product, design technology and methodology for SOC, and fabrication technology and process for SOC. From the application point of view, SOC can be divided into three classes, application specific integrated circuit (ASIC), system on programmable chip (programmable SOC), and original equipment manufacture (OEM) SOC.

With the continuous popularization of SOC applications, the market demands the prevalence of SOC design. SOC providers must not only expand their design capabilities on the interior of a system, but also directly develop and deliver the SOC design kit and methodology to their customers. Consequently, SOC design evolves toward programmable SOC.

Programmable SOC is the systematic integration, as required by products, on a field programmable chip. Multiple IC providers have introduced programmable SOC products, the system functions of which include processors, memories, and programmable logic. The most fundamental reason for the popularity of programmable SOCs is that, in addition to the avoidance of the high expense and long fabrication cycle associated with ASICs, they possess high integration with low power, small footprint, low cost of ASICs, and the low risk, flexibility, and quickness of going to market associated with field programmable gate array (FPGA).

8.2.2   SOC Samples

Many SOC products have recently come to market for various applications. Two SOC samples are cited here to give readers an idea of the application. Figure 8.2 shows a SOC sample, MXC6202xG/H/M/N, which integrates MEMS structures sensing dual axis acceleration together with signal processing circuits. Its applications include cell phones, PDAs, computer peripherals, LCD projectors, digital cameras, joysticks, electronic compass tilt corrections, etc. This SOC was fabricated on a standard, submicron complementary metal oxide semiconductor (CMOS) process and packaged with a low-profile LCC package in $(5 \times 5 \times 1.55)$mm^3. It is a complete sensor system with on-chip mixed signal processing and integrated I^{2}C bus, allowing the device to be connected directly to a microprocessor, eliminating the need for A/D converters or timing resources. It measures acceleration with a full-scale range of ± 2g and a sensitivity of 512 counts/g (G/M) or 128 counts/g (H/N) @5.0 V at 25°C. It can measure both dynamic acceleration (e.g., vibration) and static acceleration (e.g., gravity).

The device operation is based on heat transfer by natural convection and operates like other accelerometers except that its proof mass is a gas in the MEMSIC sensor.

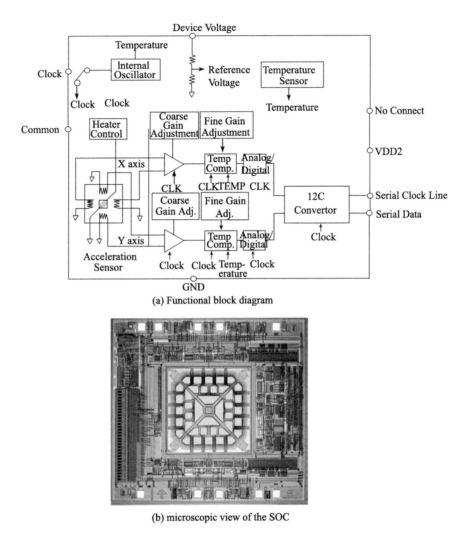

(a) Functional block diagram

(b) microscopic view of the SOC

Figure 8.2   SOC chip integrated MEMS sensing structure IC (Courtesy of MEMSIC, Wuxi, China)

The other SOC sample, as shown in Figure 8.3, is a new generation of controller system for digital TV presented by SMIC Co. Ltd. This CAM controller chip was designed with embedded high-security features. It provides a perfect secure platform for conditional access (CA) application and supports digital content protection on a common interface.

Manufactured by SMIC's 180 nm standard CMOS process technology, the SOC is embedded with ARM7TDMI, and it integrates the common interface, including the transport stream processing unit, the smart card controller, and the DES/3DES/AES/RSA cryptograph coprocessor. To address security challenges and to provide a security platform for addressing future security needs in CAM, the chip has several hardware blocks embedded to help secure the platform: security OTP block, cryptography coprocessor, memory protection unit (MPU) and secure DMA, encrypted external memory access, and security JTAG interface. It can supply good protection on software and secret information.

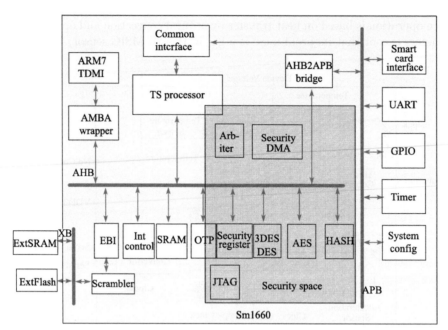

Figure 8.3   Architecture of CAM controller SOC
(Courtesy of Shenzhen State Micro Technology Co. Ltd., China)

8.2.3   Problems, Challenges, and Solutions

SOC is the ideal solution for microelectronics products, with its small size, high performance, and low cost. People hoped that SOC could evolve in a way similar to that outlined by Moore's law. However, more and more problems and challenges emerged in the development process, especially those associated with the continuous increase of the types of components needing to be integrated.

The first problem is the compatibility of semiconductor processing. With the enhancement of IC integration, the process of a purely digital IC consists of hundreds of steps. However, a typical SOC chip comprises digital units, analog units, high-voltage units, low-voltage units, radio-frequency units, and even MEMS units. The adoption of each type of unit would significantly increase the complexity and number of steps in the process. Meanwhile, the process compatibility between units of a different type is quite a concern, which embodies itself in a SOC chip with MEMS devices built on.

In general lowering costs and standardizing the process is of great concern, since the MEMS fabrication flow usually comes with specialized process steps like bulk Si etching and Si deep trench etching; the processing temperature sometimes exceeds 400°C; and a wide variety of materials are adopted. It is not a surprise that many researchers are trying to make minimal changes to an established standardized process to lower the complexity and cost, or to improve performance by modifing some of the processes for certain kinds of devices. However, these are not sufficient. The acclaimed advantages of SOC, e.g., minimized volume, ideal performance, and the lowest cost, will not be available at all if no remarkable progress is made on the process to solve the compatibility issue. Concerning the maturity of the solutions to compatibility issues, the majority of SOC products may contain nothing but digital parts, analog parts, and memories.

The second aspect is design complexity. Though both are million-gate-level chips, SOC design is obviously more difficult than that of memory. The complexity is also affected by the categories of the components integrated, and individual components need to be designed by specialized design engineers so that mask layouts of various portions can be patched together as a whole. When MEMS devices are included in particular, the process design and package design may undergo significant modifications. Meanwhile, if the package design has not been considered along with the circuit and layout design, the development may never be able to be delivered to production. Like the prediction for the of global semiconductor industry, during the first decade of the 21st century, the chip industry is predicted to reach an integration level on the order of billions of transistors, and the mainstream SOC design technologies are incapable of the full integration of a single chip on the 100-million-transistor scale.

The design complexity should be solved in two ways: on the one hand, the continuous advancement of system-level design tools is necessary; on the other hand, IP reuse techniques may be used, i.e., some programmable and general-purpose processing units. These versatile units can be adapted to the needs of various SOC chips, with programming, selection of modules, and setting of parameters. Though this will affect SOC performance to some extent, the compromise is still worthwhile when compared with the development cycle and design cost being saved. In fact, SOC design at present has shown such trends as the constant increase of IP numbers incorporated and the diminishing of the scale of customized modules.

The complexity in design also embodies itself in the design of interconnection between IP core and reusable IP cores. Currently, on-chip bus structures are used to interconnect the IP cores; that is, the cores are not connected directly to each other, rather, they are interconnected through on-chip buses. The application of on-chip buses can solve the interconnection issues of IP cores, but, the interconnection between IP cores from different vendors is barely feasible, since different vendors would use bus structures of their own, e.g. AMBA bus for ARM, EC bus for MIPS, and CoreConnect bus for IBM. Therefore, a universal on-chip bus architecture has always been the target sought after by the Virtual Socket Interface Association (VSIA). For example, an interconnection standard is proposed lately, which is in fact the data traffic between networked IP cores.

Meanwhile, the following challenges have to be faced during the design of a reusable IP core.

(1) Readability. The IP core providers must use designs described with an appropriate method, which can facilitate the use of the IP core for users and can take measures to protect intellectual property from piracy.

(2) The expandability of the design and the applicability of the process flow. An IP core should be carefully designed, verified, and optimized. An IP core needs to be of a certain range of applicability, i.e., to be applicable to a certain range of design and applications.

(3) Testability. IP cores must go through the test and verification. There still may be at least some kind of variation to the IP cores when applied to specific designs. Therefore, the functionality and performance of IP cores should also be tested by customers and should facilitate not only individual tests but also those in system application environments.

(4) The standardization of the interface definition. That is a uniform definition that should be made on the interfaces of IP cores.

(5) Copyright protection.

(6) The intactness of the data delivered to facilitate the integration procedure of chips.

(7) Low power consumption considerations.

8.3   System in Package Technology

8.3.1   Overview[3,5]

There has not been a generally accepted definition for system in package (SIP). Prof. R. R. Tummala proposed that system-level packaging can be divided into two categories: SIP and SOP.[3] That is, SIP is commonly referred to as two or more chips stacked on a single packaging substrate, which are then packaged with molding compounds and interconnected during the process of packaging. On the other had SOP may be considered the system-on-a-PCB or system-on-a-substrate, incorporating multiple chips and discrete components on a miniaturized circuit board and featuring the implementation of various system components with thin layer techniques.

In this chapter, SIP is regarded as a heterogeneous integration, including chips or modules stacked vertically to realize a three-dimentional structure; and embedded digital, RF, and optical components; and the system integrated into a miniaturized packaging system. For simplicity, let us define SIP as (multi-) functional systems built up using semiconductors and other technologies in the electronic package dimension via heterogeneous integration. SIP focuses on achieving the highest value for a single packaged microsystem. Its concept applies to quite diverse technology and application areas, ranging from sensors and actuators, RF modules for mobile communication devices, to solid-state lighting, and even health care devices, such as biosensors.

To distinguish between various SIPs, one can categorize SIP into three categories, as shown in Table 8.1.[5] SIP level 1 refers to packages with multidies, such as MCM, PiP, and PoP. SIP level 2 refers to subsystems built up using more than just the IC process, such as passive integration. The highest level, SIP level 3 refers to submicrosystems and microsystems with more than electric functions, built up using multitechnologies, such as SOP[3] and a miniature camera.

Table 8.1   SIP classification

More than...	... one IC	... IC process	...electronics
SIP level 1, e.g., MCM			
SIP level 2, e.g., passive integration			
SIP level 3, e.g., MEMS			

Generally, the SOC and SIP technologies can be complementary, not competing, approaches to achieve customer value. They are synergistic in nature, wherein SOC can be a component of SIP. The decision on which approach to use (or how to partition when both approaches are possible) is based on a thorough assessment of development and manufacturing costs and a realistic assessment of the market for the product. Rather than arguing about which is better, one should try to take into consideration the capabilities and their intrinsic advantages.

8.3.2 SIP Technology

Various companies and research institutes have designed a large variety of SIP implementations, with some pursuing the minimal package footprint, and some aiming at the minimal stacking profiles. The mainstream techniques include the following:

(1) Wire-bonding stacking integration, which has been used by early memory chips stacking SIP, has been applied to integration of the hybrid SIP of logic and memory chips. Although the wire bonding is a mature technology in the microelectronic packaging industry, it must face the challenges that come with SIP. One of the major drawbacks of the adoption of wire bonding is the significant increase of the wiring density of a packaged system containing multi-input/multi-output logic chips. As the number of layers to be stacked in SIPs continues to increase, wire bonding with various height differences is required. Consequently, while thinning down chips in SIP to a minimum, people use inverse wire bonding, i.e., they first finish the bonding on the substrate, and then implement the wire bonding on the chips.

(2) The second technique is the combination of flip-chip and wire bonding. Compared with wire bonding techniques, flip-chip can take full advantage of the die area to offer more interconnections and enhance system performance by shortening the interconnections to the system. Indeed, the complexity of SIP is also raised. On the one hand, it needs to support the flip-chip process, and on the other hand, substrate technologies with smaller feature sizes are required for high-density interconnections. The focus of the next generation of chip stacking techniques will be the connection of electrodes.[6]

(3) The third technique is package on package stacking, i.e., the replacement of chips by packaged bodies, as shown in Figure 8.4. Polyimide ribbons may be used as a flexible carrier. While acting as a thin substrate, they can be folded over to provide a mounting surface for a second overstacked layer; or rigid substrate may be used. The most significant advantages of this method are the ability to guarantee the yields by testing the individual packaged modules separately, with each module containing one or more chips. Other advantages include the overall size reduction enabled by higher wiring density on substrates, high-density

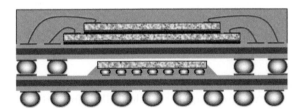

(a) Schematic diagram of whole PoP module

(b) Top view of the sample

(c) Bottom view of the sample

Figure 8.4   Schematic of package on package stacking (Courtesy of ASTRI, Hong Kong)

interconnection between packaged modules through the pin arrays, and the ability to provide a variety of packaged modules to accommodate user's requirements. The major difficulty lies in the selection of foldable substrate and interconnection on the folding edges.

(4) Substrate based integration techniques for SIP[7−15] have been studied to meet the needs of some massive applications. They include a wide variety of design and fabrication technologies, for the incorporation of sensors, RF modules, memory modules, and embedded processors with DRAM.

A traditional microsystem package takes double roles: first to provide inputs and outputs for integrated chips, and second to provide interconnections for the active and passive devices on a system circuit board.

What we are looking for is the implementation of multiple system functions in an environment that is low mass, low profile, low cost, and high performance. The design of such a system may demand high performance digital logic, memory, radio frequency, and analog signal and broadband optical performance. It does not demand a compromise in performance, since the individual technologies are incorporated discretely. In addition, the complexity of system design and the time requirement are reduced, so the testing becomes easier.

Generally, this system integration technology is an advanced system-level packaging technology, involving the whole set of system functions and interconnections. In a sense, it assimilates merits of SOC, MCM, 3D-stacking, etc., to accomplish system integration on the highest system level and in the most cost effective way. A typical sample was implemented by the Georgia Institute of Technology, which shows the first integration example of optical, RF, and digital functionalities in a single module[8] intended for the broadband application of a smart network switcher.

(5) Through silicon via (TSV) has been widely accepted as the one of the techniques enabling the effective and highly dense three-dimensional integration of hetero- or homogeneous ICs and micromachined functional structures, such as sensors, actuators, and antennas, in a package. This type of vertical interconnection may provide a short and low-resistance/impedance interlayer signal path for these stacked chips or modules, so that the relatively high signal loss, temporal delay, and noise pickup associated with the long interconnections in the in-plane integration methodologies can be mostly eliminated. In addition, TSV may provide the flexibility in 3D interconnection wiring and layout for both digital and RF/microwave circuitry.

Currently, two generic approaches for 3D SIP fabrication are used, i.e., "vias first" and "vias last" approach. In the "vias first" approach, the TSV is formed before the standardized fabrication processing of IC wafers. In the "vias last" approach vias are formed after the IC is fabricated. The "via last" approach is selected in this work, since no harmful impurities (from the IC process point of view) are introduced, and the process is relatively simple. However, precise control of post-IC process temperature and careful selection of etching agents are needed for the integrity of the IC fabricated.

The integration of the 3D SIP based on TSV involves several important processing steps: formation of vias; deposition of via insulation, barrier, and seed layer; via filling; wafer thinning; wafer stacking; and so on, as shown in Figure 8.5.

The formation, i.e., the drilling, of the vias is one of the critical process steps for the success in the TSV microfabrication. Among the technical options, the deep reactive ion etching (DRIE) is considered to be the ideal one for batch drilling of deep vias into Si substrates.

In order to insulate the subsequent via metal from the surrounding silicon, SiO_2 is deposited by plasma enhanced chemical vapor deposition. Ideally, coverage over the entire via structure should be conformal. However, the relatively high aspect ratio of the via can pose a

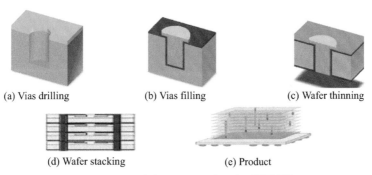

(a) Vias drilling   (b) Vias filling   (c) Wafer thinning

(d) Wafer stacking   (e) Product

Figure 8.5   A key process flow for TSV SIP

problem requiring careful optimization of deposition parameters. It is also critical to control stress in the insulating layer, in order to ensure structural integrity under subsequent processing and operating conditions. The titanium-tungsten (TiW) barrier and the Cu seed layer are deposited by magnetic sputtering. There are two methods of copper electroplating. One is DC plating, the other is periodic pulse reverse (PPR) current plating. In TSV's copper electroplating, cupric ions are reduced to copper and deposited on the cathode, but the deposition rate is different between the top of the via and the bottom of it. Because the depletion rate of cupric ions is faster than the diffusion rate of cupric ions from the bulk solution into the bottom of vias, the concentration of cupric ions at the entrance of vias is higher than that at the bottom. This concentration gradient leads to different deposition rates: the deposition rate at the top is faster than that at the bottom. Eventually, voids or seams are usually formed inside of vias. In the case of DC plating, there was only one nucleation step, and copper grains grow continuously with electroplating time, since it provides a continuous current with a constant current density and the via is always kept at the cathode potential. So, it is difficult to obtained defect-free copper filling for vias by DC plating.

The next step is wafer thinning. The packages of products such as mobile phones are thinner and thinner. The maximum thickness required is currently 1.2 mm, and the thickness of packages of some products has even reached 1.0 mm. As the number of stacked chips increases, the thickness of chips must decrease. There are three main methods for thinning chips—mechanical grinding, chemical etching, and atmosphere downstream plasma (ADP). The mechanical grinding method can usually reduce the thickness to about 150 μm, while the plasma etching method can reduce the thickness to 50–100 μm. A method for reducing the thickness to 10–30 μm is being developed. In addition, handling thinner wafers with thicknesses smaller than 50 μm is also a big problem. A glass carrier wafer is often used to temporarily bond the device wafer in the postthinning processes. [16]

Wafer bonding or wafer stacking is a key process to form a 3D structure by attaching chip-to-chip (C2C), using metallic bonding, dielectric bonding, or metallic/dielectric hybrid bonding. Metallic bonding with low-temperature solder and metallic bonding with Cu-Cu thermal compression are widely used. Wafers or chips can be connected face-to-face (F2F) or face-to-back (F2B). Memory stacks tend to use F2B stacking, while memory-logic prefers F2F stacking. [16]

In addition to the manufacturing technique issues, it is critical to select an integration approach as well, especially to decide whether a complete chip stacking or a packaged module stacking architecture should be used. SIP based on packaged module stacking has a more appealing manufacturing cost. Alternatively, a pure chip stacking can be more advantageous, if the yield of individual chips is high enough, as shown in Figure 8.6. In fact, the SIP methodologies in practice can be flexible and diversified and can be a combination of various integration techniques.

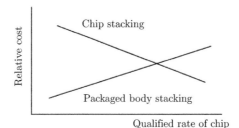

Figure 8.6   Manufacturing cost vs. integration methods

Currently, stacking technology faces three major challenges. The first lies in microint-erconnection. With the shrinking of package and the increase in the number of chip in-put/output ports, the method used to construct the interconnection between a very large numbers of pins with extremely fine pitch becomes crucial. The relevant SIP package sub-strate needs to carry ultrafine wiring, and more refined layer-up substrate has to be built to accommodate the requirements of higher speed and density. The second challenge is stacking chips. SIP technology can provide stacking up to nine layers, e.g., the packaged five-chip stack-up from Renesas Technology has a profile of 1.5 mm, with a 1 mm packaged profile targeted for its next generation stacking SIP with over 10 layers. However, with the increase in the number of stacked layers, how to build the stacking with thinner dies and to guarantee the reliability becomes quite a concern. At the same time, carrying out efficient heat dissipation and cooling is required. A particularly severe problem is how to test the tier up to ensure the proper operation of chips and the yield of SIP products. The third challenge is the adaptation to the SIP design environment. Currently, the design software environ-ments for SIP relatively fall behind and demand improvement and enhancement based on the inherent design resources. In particular, as the march into high-frequency bands has increased the ratio between line length and wavelength more and more, more sophisticated distribution parametric models should be to replace used the traditional lumped model in analysis and design; in addition, new issues will arise from the adoption of SIP, such as the electromagnetic radiation coming along with the high-speed signal transmission. In order to realize a genuinely global optimization, efforts on these issues are indispensable.

8.3.3   Comparison between System-on-chip and System-in-package

As we have mentioned previously, diversified integration methodologies, such as System-on-chip(SOC), SIP, and System-on-chip(SOC), each has advantages and disadvantages, and one can find their respective role in real applications. Furthermore, during the process of continuous development and advancement, breakthroughs in one technology can alter an applications status quo. Based on the current technology, we can discuss preliminarily the advantages and disadvantages of these system integration methods.

When choosing a system integration mode, people need to consider a specific product design from the point of view of product features, development cycles and market scale. In terms of the integration of heterogeneous devices, SIP can conveniently realize the integration of analog and logic circuits, CMOS and SiGe, LSI, and various passives. From the cost and profit point of view, SOC is advantageous for large volume production, and SIP is advantageous for small volume production. From the reuse of resources, SIP may incorporate off-the-shelf chips, SOC can reuse existing IP; however, as a whole, SOC design is more sophisticated and difficult. From the design point of view, SOC design is made from a specific technology library, and SIP may reduce the number of components and the size and layout complexity of PCB; however since there is a lack of research on corresponding design fundamentals, in many cases, vendors have to do the technical research themselves.

Compared with SOC, the SIP approach has overcome its technical limitations, especially in the global wiring time-delay and RF integration aspects. The size of the SIP wires is much larger than that of SOC, which facilitates internal interconnection of high quality. Optical transmission can be used to replace traditional copper traces as the high-speed interconnection between chips and other components. This not only mitigates the ever rising wire resistance problem, it also avords signal cross talk. Furthermore, the limitation of RF integration inherent to SOC can be overcome through SIP. It is more appropriate for RF devices, such as capacitors, filters, antennas, transducers. and high-frequency high Q inductors to be integrated through packaging than monolithic integration.

The system-on-chip is the traditional integration method, and comparatively, the system on package has just emerged. However, considering the rapid developments of recent years, it is reasonable to believe that, in the coming years, these two technologies will make significant progress in wider applications.

8.4   RF System Package Technology

8.4.1   Overview

Radio frequency, in a broader sense, covers the electromagnetic spectrum, ranging from the low end of long-wave radiotion (30 kHz) up to below the low end of the far infrared-band (400 GHz). In recent years, the trend for RF technology is miniaturization, reduction of power consumption, and improvement in reliability and multiple functions. [17−32] In addition, they demand filter devices with high-frequency selectivity and local oscillators of high stability,[29,33,34] and need to ensure the circuits can operate for a long time with limited power supplies (such as chemical and solar cells). Thus, the RF circuit should be further integrated. The capacitor, inductor, and oscillator should have a high Q value (quality factor); the transmission line, switch, and antenna must meet higher requirements in impedance matching, insertion, and isolation aspects. [35]

In a typical wireless transceiver front-end, passive devices, such as antennas, transmit/receive switches, mirror suppression filters, bandwidth filters, crystal oscillators, intermediate filters, inductors and capacitors, are at present in discrete form, and can only be interfaced to active devices on board level, occupying 80% of a circuit board. [36] On the other hand, on-chip silicon or compound semiconductor devices are limited by factors such as material properties and operation mechanisms, and thus their performances, e.g., Q factor, isolation, insertion loss, control power consumption, etc., are difficult to improve. These factors have become obstacles to the advancement of the RF system. MEMS technology provides new approaches for solving these problems. Since 1990, many research organizations from all over the world reported novel RF devices fabricated with this technology. Those devices have demonstrated performances that are superior to those of traditional on-chip devices, even exceeding those of discrete components, and possess monolithic integratability. They can not only replace existing components, but promise to alter fundamentally traditional transceiver architecture, [29,30] as shown in Figure 8.7. The merging of MEMS and RF technologies gave birth to the new concept of radio frequency micro electromechanical system (RF MEMS). Currently, it covers the various miniaturized and integratable components intended for signal generation and process for frequencies ranging from common radiowaves up to the microwave/millimeter wave band, and relevant design, fabrication, and package technologies.

Investigation of RF MEMS started in the early 1990s, the tag of which was the high performance rotary MEMS switch from Hughes Research Labs. [17] Package issues were not considered at that time. With the RF MEMS devices and circuits going on the market, people gradually realized that packaging is one of the supporting technologies enabling RF MEMS to really display their advantages, and the pivotal role of packaging technologies is increasingly prominent. [18−20] Compared with the device research, current investigations in RF MEMS packaging are still not mature.

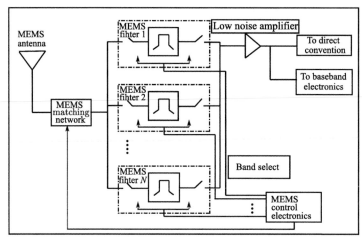

Figure 8.7   A novel RF MEMS-based RF transceiver architecture

1. Device Categorization

(1) On/off and channel selection devices. These devices are mainly various switches and relays; they hold the largest share of the currently investigated RF MEMS devices, and the relevant principles, design, and fabrication technologies are comparatively mature.[35−48] According to the control mode of the RF signal, they can be further divided into shunt and series devices. The former rely the shunt/reflection of signal to implement control of the signal channel, suitable for the on/off control in the high-frequency range (5−100 GHz); the principle of the latter is very similar to traditional DC relay and low-frequency switches and is suitable for the lower frequency range (0.1−40 GHz). Two typical examples are displayed in Figure 8.8. At present the RF performances of micromachined switches on the whole surpasses those of field effect transistor (FET) and PIN devices (particularly in insertion loss, isolation, and IP 3 performances), and their switching time and power handling capability barely surpass the two semiconductor devices.

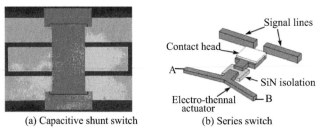

(a) Capacitive shunt switch   (b) Series switch

Figure 8.8   Micromachined switch examples developed in Beijing University

(2) Lumped energy storage and frequency selection elements. Various tunable capacitors and inductors are included. Currently, these devices are for the frequency selection circuit and tunable filter circuit, whose Q factor determines the frequency selectivity and insertion loss performances of circuits. Lumped elements are of great value in high-power oscillation, power amplification, and wideband circuit, and they are most suitable for monolithic microwave integration circuits, which require elements with minimum footprints.[49]

(3) Electromechanical resonators/filters. In reference frequency signal generation and channel selection, thin film bulk acoustic and surface resonant structure resonator/filters are expected to replace traditional quartz crystal oscillator and surface acoustic wave devices, which are hard to integrate on the chip. The micromachined bulk acoustic devices are based on piezoelectric thin films, with Q factors over 1000, which is suitable for high-frequency applications in ultra high frequency and super high frequency bands, and difficult to trim.[14]

(4) Micromachined transmission line, waveguide, millimeter wave resonator/filter, and antenna. [50−55] These devices transmit/receive signals in microwave bands or above including i) Planar transmission lines. Micromachining is used to remove the substrate of high dielectric constant to reduce the loss introduced by the substrate and the dispersion/non-Transmission electron microscopy mode associated with the substrate/air interface. In addition, the shielded cavity may be micromachined to reduce interference and radiation losses. Thus, high-performance high-frequency transmission lines may be manufactured on a low resistive substrate. ii) As shown in Figure 8.9(a), a micro-waveguide operates in the hundreds of GHz frequency range, where planar transmission lines have poor performances, and traditional waveguide is hard to fabricate. LIGA or UVLIGA can be used to form micro waveguides that can be integrated on chip. iii) RF resonator/filter which demands a precisely distributed structure, is shown in Figure 8.9(b), Micromachining may implement a resonating transmission line structure with high dimensional accuracy. The Q value is comparable to a conventional RF resonator and filter. iv) Antenna, as shown in Figure 8.9(c). MEMS techniques may be applied in three ways. The first is to fabricate a planar antenna in the millimeter wave or higher frequency range on a dielectric membrane to reduce the effect of substrate on the radiation efficiency. The second is to fabricate a horn-like reflector or actuator to enhance the radiation directivity or tune the radiation direction. The third is to etch the dielectric membrane underneath the antenna into an aperture array with a certain pattern to tune the dielectric constant, to enhance the antenna efficiency, and eventually to realize the monolithic integration of the antenna on a high-dielectric constant RF circuit substrate.

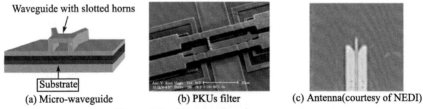

(a) Micro-waveguide   (b) PKUs filter   (c) Antenna(courtesy of NEDI)

Figure 8.9   Micromachined components

(5) Passive circuit functional modules through a combination of multiple RF MEMS devices. Combining multiple RF MEMS components like switches, capacitors, inductors, etc., together, one can build some relatively independent passive circuit units with specific functions, [35,37,38,55] such as a single-pole-multi-throw switch network, a tunable filter network, a passive mixer, a phase shifter, a tunable matching network, a programmable attenuator, a high-performance switch network, etc. Figure 8.10 shows a compact five-bit digital MEMS phase shifter with three-port RF MEMS switches. It is based on the switched-line design and 20 switches with an insertion loss of 0.66 dB and an isolation of 23.3 dB at 10 GHz. This phase shifter is fabricated on high-resistivity silicon substrates that are 7 mm×4 mm×0.2 mm. The phase shifter performed well in the X band, with an average 3.6 dB insertion loss and phase shift error less than 5° at 10 GHz.

The emergence of these comparatively independent functional modules may elevate the integration level of the whole circuit, simplify their structure and system architecture, and lower the loss and noise, which is a noteworthy development trend.

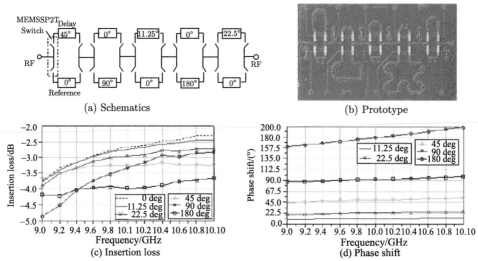

Figure 8.10   A five-bit MEMS phase shifter (Courtesy of NEDI, China)

2. Features and Prospects

Compared with conventional RF technologies, RF MEMS demonstrates the following prominent advantages: (1) excellent RF performance, such as low insertion loss, high isolation and Q factor, and outstanding nonlinear properties; (2) low driving power consumption; (3) convenience of integration with active circuits; and (4) potential of massive and highly precise production making use of existing mature IC process. Hence, RF MEMS possesses a bright future, which may not only replace existing passive and modules in traditional architecture-based RF circuits and lead to an all-in-one transceiver with outstanding performance, but may also result in a novel, more optimized RF transceiver system. However, current RF MEMS devicesuhave drawbacks, such as the slow switching action, the limited power handling capability, and the lack of reliability of the movable structure, which does not surpass existing devices.

With the development of these technologies, companies like Radant MEMS started to provide customers with switch samples for trial.[57−60] Various MEMS market analysis organizations gave relatively optimistic estimations for applications and market prospects in the coming years. WTC (Wicht Technologies Consulting), a consulting company, carried out an in depth analysis of RF MEMS technology development during 2002−2007. This company believed that the RF MEMS market could keep a strong growth momentum, and its scale would amount to over 1 billion U.S. dollars in 2007.[61]

8.4.2   The Features and Roles of RF MEMS Packages

1. Features of RF MEMS Packages

RF MEMS packages encounter common questions in MEMS and Microsystems packages, but they also have to address specificity and complexity, as discussed below.

(1) RF MEMS involves multiple disciplines, and its fabrication, materials, and mechanisms reveals diversity, which is a challenge to the knowledge structure and technology integration capability.

(2) MEMS fabrication technology is unique in its usage of materials and feature size of fabrication, which may conflict with the packaging and fabrication technology, and the use

of high-performance transmission lines, matching structures, and interconnection of vias in RF IC will be limited.

(3) RF MEMS packages usually couple with internal interconnection transmission lines and microstructures, e.g., in a packaged micromachined switch, the analysis and simulation of RF characteristics of the switch bridge membrane of the coplanar waveguide (CPW) need to involve the effect of package shield. [62] Consequently, package design for RF MEMS and the integration of RF MEMS devices and circuits are all challenges package technology R&D personnel must confront; the consideration of package and integration technologies must penetrate the whole process of the development of RF MEMS devices and circuits, otherwise it would be difficult to design usable devices. Conversely, the advancement in RF MEMS fabrication technologies and package technologies will surely supply more original and optimized package forms for RF IC and monolithic microwave integrated circuit MMICs, and high-frequency/high-speed MCM.

Technically, the RF MEMS package is the interface between internal devices and external circuits or systems, and it must provide protection and interconnection. On the one hand, it needs to ensure the mechanical actuation property of the micromachined structure inside; on the other hand, it should guarantee the normal exchange of DC and high-frequency signal while minimizing the signal loss and suppressing the interference, noise, and electromagnetic radiation. Furthermore, the package structure itself should be simplified as far as possible, in order to meet the massive production requirements; the cost and weight should also be minimized.

2. Role of RF MEMS

(1) Protection of internal dies. Following are the protective functions of the RF MEMS package. On the one hand, it must protect the internal sensitive micromechanical structure from interference and destruction by external erosive factors and environmental factors, e.g., vibration, heat, dust, humidity, electromagnetic interference, polluted or erosive atmosphere, etc. If no protection is provided, the chip leads would be damaged and various movable structures could get stuck, and the reliability of the device would be affected. On the other hand, the internal atmosphere of the RF MEMS package should be able to ensure the dynamic property of micromechanical structure movement, and the packaging material themselves shald not introduce ingredients harmful to the dies. In highly demanding applications, RF MEMS usually requires the use of hermetic packaging, or even vacuum packaging, and materials with minimum air adsorptions and release for normal operation of internal devices.

(2) Interface. The package is expected to capable of relaying DC and high-frequency signals. The interconnection of the high frequency signals inside and outside the package may be implemented with traverse metal transmission lines, coaxial cables, and vertical vias. In this case, a designer must pay close attention to the minimization of the impact by reflection and loss due to the transmission mode difference between internal and external signal traffic, which is inevitable. Three dimensional finite element or method-of-momentum simulation software, such as Ansoft HFSS, Microwave Office, Momentum, is a good high-frequency characteristics analysis tool. [63]

(3) Mechanical support. The physical properties of the mechanical support and molding materials for packaging should be as close to that of the die substrate as possible, in case large thermal stresses appear between two interfaced materials when the temperature varies and induces cleavering and delamination.

(4) Thermal conductive dissipation. When a device needs to handle high power signals, temperature elevation cannot be neglected. In this case, the thermal management of the device should be considered carefully, and materials with good thermal conductivities should

be used.

8.4.3   Categorization of RF MEMS Package[64−73]

The package design of RF MEMS depends on the requirements of intended applications, system structures, and optimization. Currently, the packaging methods for RF MEMS are diversified, include all metal, ceramic, plastic, multilayer, wafer-level packages, and microshield and self packages, the first three of which derive from the conventional semiconductor package, discussed in detail in Chapter 5.

1. Metal Package

The most notable advantage of metal packages is that they offer the best hermetic sealing. In addition, the metal housing possesses excellent heat dissipation and electromagnetic shielding capability. In the RF system-level package, metal packaging was first applied to MMICs and hybrid circuits. For example, CuW (10/90), Ag (Ni-Fe), CuMo (15/85) and CuW (15/85), and Kovar alloys offer excellent heat conductivity, superior to that of Si. These metals, combined with Cu, Au, or Ag plated transmission line, may enable great RF MEMS package.[54] During the design of a metal package, it is important to consider the effect of metal housing on internal signal transmission lines and micromechanical structures. For example, the CPW (coplanar waveguide) often takes the shape of the shielded coplanar waveguide, whose transmission characteristics such as wave speed would be quite different from typical CPWs.

2. Ceramic Package

Ceramics come with the advantages of low weight, low cost, and ease of mass production, and thus have become one of the most frequently used packaging forms in microelectronics. Ceramic packages can be made hermetically, support multilayer design, and facilitate the integration of multilayer circuits through signal feeds. The multitier package on the one hand reduces the overall size of multiple MEMS chips put together, lowers the cost, and may shorten the distance of interconnection, which is beneficial for the optimization of the package interior. The integration with ceramic stackup and interconnection may change the RF chracteristics of the package.

In a high temperature cofired ceramic (HTCC) package, the cofiring temperature may reach 1600°C. The most frequently used metal sure is tungsten (W) and molybdenum (Mo) with relatively low electric conductivity. In the low-temperature cofired ceramic (LTCC) package, the cofired temperature is below 1000°C, so conductors like silver (Ag), gold (Au), and Au/platinum (Pt) may be used, and the passive components such as resistors and capacitors may be embedded to the LTCC stacks, so that the wiring length may be reduced. The problem associated with ceramic packages is the shrinkage of the green tape during the baking process. Thus, iterated simulation and experiments are necessary in the wiring. Furthermore, the bonding strength between the ceramic and metal is inferior to that between ceramics. The processing temperature of ceramic limits the available metal selection, and it is important to ensure that no reaction occurs between the metal and the ceramic.

3. Plastic Packaging

Plastic packaging is popular in the microelectronics industry due to its low cost. However, it cannot offer the hermeticity demanded by high-performance applications. Plastic packaging tends to crack in the process of temperature cycling tests, and the gas released in the ramp up process may affect the physical behavior of devices. In addition, plastic packaging is formed with transfer molding, so cavities for holding micromachined structures are hard to form. Therefore, new RF system packaging methods with better sealing performance capable of forming cavities are under development. [64,65]

4. Multilayer Stack Package

A 3D multilayer package based on MEMS structures was reported. [66] Passive components such as filters and corresponding circuits may be formed on each layer, and active components are mounted on the top layer with flip-chip technology. This 3D hybrid circuit structure uses Si instead of salliom-arsenic (GaAs), and then the Si-based micromachining techniques may be used to form the 3D interconnection structure.

Embedded overlay is considered to be one of the multilayer packages. This technology is suitable for microop to electromechanical system (MOEMS) and RF MEMS packaging. Formally, this type of package looks like the upside-down form of an ordinary multilayer package. The concept of embedded overlay derives from the chip-on-flex (COF) in microelectronics packaging domain. Packaging involves encapsulating chips with molding plastic substrate and forming the interconnection through the patterned metal thin film on the coating layer covering the module. [68] The chips are attached onto COF film with polyimide or thermal-plastic adhesives. Conversion, compressive molding, or transfer molding may be used for encapsulation to mold the chips and module together at 210°C, and at the same time to form a carrier. Vias can be ablated with a continuous argon (Ar) laser to form vertical interconnections. Variable laser ablation power level, combined with process such as plasma cleaning and high-pressure water flushing, may remove the unwanted COF cover layer while not damaging the embedded devices. Currently, this packaging process may only provide single layer interconnection, but it is progressing toward multilayer interconnections.

5. Wafer-level Package

Designers should take packaging into consideration in the very early stages of RF MEMS devices R&D. The research on cost-effective packaging techniques is focusing on wafer-level packaging.[67−69] If the packaging method is defined in advance and merged into the device development process flow, the cost is bound to be reduced and the optimal package structure will be obtained. Figure 8.11 shows a schematic of wafer-level packaging. Since movable parts are on the surface of the RF MEMS silicon substrate wafer, the bonding of the device substrate wafer and a silicon or glass wafer with a large amount of capping (or lid) structure micromachined will greatly enhance the performance of the device and expand its range of applications. These lid-like structures are formed with dry or wet etching, corresponding to the individual devices. The bonding may be done with methods such as anodic bonding and frit glass; the lid-like structure provides a vacuum or inert gas atmosphere, to avoid contaminating the structure. Furthermore, wafer-level packaging can provide protection during the final dicing and splitting of the chips, shielding them from damage by high flow rate water jets in the dicing and from dust contamination.

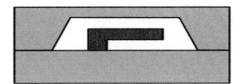

Figure 8.11   Wafer-level package for RF MEMS

Wafer-level packaging should be taken into consideration in the initial stages of the design of a product. The lid structure may be accomplished by bulk etching cavities into Si wafers or glass wafers. On Si wafers, the Si lid structure may be accomplished by bulk silicon anisotropic wet etching and ICP deep silicon etching. The silicon lid obtained by wet etching usually takes the shape of an upside-down pyramid. The reason for this is that the etching along the $< 1\ 0\ 0 >$ and $< 1\ 1\ 0 >$ crystal planes progresses at the highest rate, while the etching rate along the $< 1\ 1\ 1 >$ plane is the lowest. Typical mask layers include silicon dioxide, LPCVD SiN, or the composite of both. Deep dry etching can create a cavity with an arbitrary planar opening shape and steep side walls.

6. Microshield and Self-Packaging[55,70]

Micromachining is not only used to fabricate high-frequency transmission lines with low loss and dispersion characteristics by removing part of the substrate, it is also used to form capping structure, to protect the micromachined high-frequency transmission lines against external electromagnetic interference, an example of which is the microshield and self-packaging method shown in Figure 8.12. An air-filled cavity is etched into the upper silicon wafer, and a self-packaged circuit is formed by the bonding of upper and lower shield circuits.

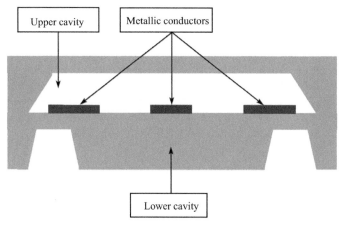

Figure 8.12   Self-packaged circuit constructed out of two silicon wafers

8.4.4   Materials for RF MEMS Packaging

One key guideline for the selection of materials for RF packaging is to ensure low signal loss. RF electromagnetic fields, when propagating along transmission lines, will interact with the conductor and dielectric constituting the transmission line, so there exist both conductive losses and dielectric losses. These losses may be avoided. In addition to dielectrics, RF package also needs highly conductive conductors. Tables 8.2 and Table 8.3 list typical dielectrics and conductors. [67,71]

Table 8.2   Dielectrics for RF packaging

Dielectric	Dielectric constant	Thermal coefficient of expansion $(10^{-6}/(^\circ C))$	Thermal conductivity $(W/m \cdot K)$	Fabrication temperature $^\circ C$
92% Al$_2$O$_3$	9.2	6.0	18	1500
86% Al$_2$O$_3$	9.4	6.6	20	1600
Si$_3$N$_4$	7	2.3	30	1600
SiC	42	3.7	270	2000
AlN	8.8	3.3	230	1900
BeO	6.8	6.8	240	2000
Resin-Kevlar	3.6	6.0	0.2	200

Table 8.3   **Typical conductive materials for RF packaging**

Metal	Melting point (°C)	Resistivity $(10^{-6}\Omega/\text{cm})$	Thermal coefficient of expansion $(10^{-6}/°\text{C})$	Thermal conductivity $(\text{W/m} \cdot \text{K})$
Au	1063	2.2	14.2	397
Al	660	4.3	23.0	240
W	3415	5.5	4.5	200
Cu	1063	1.7	17.0	393
Ag	960	1.6	19.7	418
Kovar alloy	1450	50	5.3	17
Au–20%Sn	280	16	15.9	57

Typical RF circuit substrates have one disadvantage. They only allow the conductor traces to propagate along their upper or lower surface. However, the advancement in cofired ceramics enables the conductor trace to propagate in between layers. Cofired ceramics can be divided into high-temperature cofired ceramic (HTCC) and LTCC, as discussed in Chapter 4, section 4.5 HTCC material is fired at 1600°C, and then refractory metals like W should be used as wiring material.

8.4.5   Chip-to-package and On-chip Interconnection

RF chip-to-package interconnection is one key issue. The three fundamental interconnection methods are usually adopted as the RF system interconnection: wire bonding, flip-chip bonding, and tape automated bonding (TAB).

Wire bonding is the most widely used interconnection method. Metal interconnecting wires such as Au are connected to the bonding pad on the die and then connected to the pads on the package. For RF applications, wire bonding presents two problems. The first is that the electrical resistance of the thin wire and the bonding joint is high enough to induce significant loss. The second is the difficulty in producing balanced transmission lines. The wire bonding introduces parasitic inductance associated with the wire length. The arc shape at the interconnection bonding joints may cause a certain inductance. For equipment demanding a large number of interconnections, Au-plated copper should be used to replace pure copper wires. Multifold wires may be used to accommodate high-current applications, however at the cost of higher parasitic inductance.

TAB bonding is an extension of the wire bonding concept. One advantage of TAB is that the geometric sizes of the interconnection beam thus formed are more deterministic than that obtained with wire bonding, so that better transmission lines may be formed. Meanwhile, as the interconnecting wire length becomes shorter, the inductance is accordingly reduced. Flip-chip bonding refers to the direct attachment of a flipped die onto a substrate, which enables direct contact between pads on the die and substrate[55,71]. In a typical IC package, epoxy is used to underfill the space between chip and substrates, protecting the die against humidity, and lowering the thermally induced stress. Obviously, those fillers are unfit for most RF MEMS dies. The connection of the flip-chip process has the highest connecting density among standardized processes. For RF systems, it greatly shortens the transmission line length to reduce parasitic inductance and capacitance.

Usually on-chip interconnection does not fall into package issues. However, RF MEMS and MMIC usually use special interconnection structures. Thus, the on-chip interconnection, such as the material selection and transmission line matching, is closely related to packaging. In addition, micromachining may optimize the structure of transmission lines and elevate the integration, and the use of this technology must take into consideration compatibility with package-level interconnection.

RF MEMS and MMIC generally use the planar transmission line or waveguide as internal high-frequency signal interconnections. The behavior of these transmission structures depends on frequency. Their geometry and material properties determine their characteristic impedance and loss, and these two parameters conversely affect matching with other modules and transmission lines. The RF current of a transmission line flows along the conductors, and the electromagnetic field propagates in the surrounding dielectric. The conductivities of the materials always exist, so conductive loss and dielectric loss are induced. Moreover, transmission lines of specific sizes will cause radiation loss. The complexity of RF transmission line design lies in minimizing the wire loss and simultaneously obtaining suitable characteristic impedance. Figure 8.13 is cross-sectional views of various waveguides. Table 8.4 lists their basic properties.

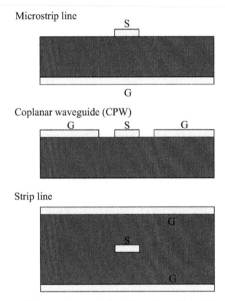

Figure 8.13   Three types of planar transmission lines, in which S denotes the signal line and G denotes the ground

Table 8.4   Planar waveguide characteristics

Characteristics	Microstrip	Coplanar waveguide	Strip line
Radiation loss	Low	Low	Very low
Unloaded Q	Medium	High	Low
Power handling capability	Medium	Low	Medium
Fabrication	Double-layer metallization	Coplanar	Embedded in between conductors
Feature	Grounding	Easy to fabricate	Connection to the mid conductor

Planar transmission lines are usually compatible with IC technologies and with RF MEMS processes as well. There are three basic planar waveguide types: microstrip, coplanar waveguide (CPW), and strip line. Microstrip line may be conveniently fabricated with IC or MEMS processes, and it is the most compact among the three. Ideally, its ground plane is infinite. However, the board sizes must be finite. Thus, there must be a tradeoff between performance and compactness. One problem faced by microstrip is that the line width of a signal line is always limited by the characteristic impedance required and dielectric thickness. Usually, the ground plane is made on the backside of the chip, which means that the die thickness also limits the dielectric thickness. A thinner backplane may lower the dielectric loss but demand a narrow signal line, and this conversely increases the conductor loss. One solution is to selectively thin the area underneath the signal line to reduce the loss at the cost of elevated complexity of process. An extreme method is to use thin film to support the signal line, but another problem emerges because of the separate placement of the signal line and ground line on the double sides of the backplanes: it is difficult to connect the microstrip to other modules or transmission lines. Precision machining may be used to drill vias to solve this problem but may make it more difficult to fabricate and increase the cost.

The strip line has the advantage of better shielding, but at the cost of more complex fabrication. The strip line is rarely used in RF MEMS because its conductor is difficult to access. The major CPW feature is that they are truly planarized, which means that they are easy to be fabricated, measured, and wire bonded. Meanwhile, the thickness of the backplane has little impact on the characteristics. The characteristic impedance depends on the line width of the signal and ground lines, the spacing between signal and ground, and the effective dielectric constant. Analysis of the planar transmission line must account for the existence of package housing, e.g., the CPW should be analyzed as shielded CPW. The analytical analysis methodology may be found in Liu. [72] Researchers use various materials or novel transmission lines to reduce the loss in RF MEMS and increase the isolation. One may refer to Clark et al. [55] for the corresponding methodology.

8.4.6   Hot Topics in RF Microsystems Packaging

The great challenge of RF microsystems packaging is the system optimization over a wide range of frequencies. Table 8.5 lists the various challenges in RF microsystems packaging and their potential solutions. [54,55]

Table 8.5   Challenges faced by RF MEMS packaging and potential solutions

Packaging steps	Challenges	Potential solutions
Release and dry-up	Stress in the devices	Dry freezing; supercritical CO_2 dry-up; roughen the contact surface, such as bumping or automatic adhesive coating
Wafer dicing and splitting	Risk of contamination, elimination of particles	Release die after dicing; laser dicing; Wafer-level hermetic packaging
Chip handling	Device failure, the sensitivity of the top chip to the contact	Use fixture to clip the rim of MEMS chip, instead of fixing from the upper surface
Stress	Performance degradation, variation in resonant frequency	Use adhesive with low elastic modulus for die attachment; annealing; use materials with matching CTE
Gas release	Stiction, erosion	Use epoxy of low gas release, cyanate, ester, low modulus solder; new adhesive for die attachment, to remove the gas released
Test	Nonelectrical stimulation	Use all kinds of wafer-level probing methods available to test; if necessary use a testing system with high performance-cost ratio

Questions

(1) What are the features of SOC and SIP? What are the pros and cons of each type?

(2) Describe briefly the features and role of RF system packaging. What are the differences between the interconnections in RF system packaging and the traditional interconnections?

(3) Please describe how to test and evaluate the quality and function of the chips inside a system package.

(4) Please explain how to realize the thermal management of a 3D system package.

References

[1] Shuidi Wang, Jian Di, and Songliang Jia. System on a Chip and System In a Package. China Integrated Circuit, 4 (2003): 49–52.

[2] Minbo Tian. Electronic Package Engineering. Beijing: Tsinghua University Press, 2003

[3] Tummala, R.R. Fundamentals of Microsystems Packaging. New York: McGraw-Hill, 2001.

[4] Rickert, P. and Haroun, B. SoC integration in deep submicron CMOS. IEEE International Electron Devices Meeting, (2004): 653–656.

[5] G.Q. Zhang, F. van Roosmalen, M. Graef. The Paradigm of More than Moore. In: Proceedings of the 6th International Conference on Electronics Packaging Technology, Shenzhen, China, Aug. 30 to Sep. 2005.

[6] Pienimaa, S.K., Miettinen, J., and Ristolainen, E. Stacked modular package. IEEE Transactions on Advanced Packaging, 27.3 (2004): 461–466.

[7] McCaffrey, B. Exploring the challenges in creating a high-quality mainstream design solution for system-in-package (SiP) design. In: 6th International Symposium on Quality of Electronic Design, 2005: 556–561.

[8] Tummala, R.R. SOP: what is it and why? A new microsystem-integration technology paradigm-Moore's law for system integration of miniaturized convergent systems of the next decade. IEEE Transactions on Advanced Packaging, 27.2 (2004): 241–249.

[9] Biye Wang. Considerations for MEMS packaging. In: Proceedings of the Sixth IEEE CPMT Conference. (2004): 160–163.

[10] Lee, Y.C., Amir Parviz, B., Chiou, J.A., et al. Packaging for microelectromechanical and nanoelectromechanical systems. IEEE Transactions on Advanced Packaging, 26.3 (2003): 217–226.

[11] Guo Wei Xiao, Chan, P.C.H., Teng, A., et al. A pressure sensor using flip-chip on low-cost flexible substrate. Electronic Components and Technology Conference, (2001): 750–754.

[12] Khalil Najafi. Micropackaging Technologies for Integrated Microsystems: Applications to MEMS and MOEMS. Micromachining and Microfabrication Process Technology, 3.4979 (2003): 200.

[13] http://www.shellcase.com/pages/products-shellOP-process.asp#top downloaded Oct. 7th, 2005

[14] Okamoto, K., Morie, T, Yamamoto, A., et al. A fully integrated 0.13 μm CMOS mixed-signal SoC for DVD player applications. Solid-State Circuits, 38.11 (2003): 1981–1991.

[15] Keilman, J.R., Jullien, G.A., and Kaler, K.V.I.S. A SoC bio-analysis platform for real-time biological cell analysis-on-a-chip. In: The 3rd IEEE International Workshop on System-on-Chip for Real-Time Applications, (2003): 362–368.

[16] Roy Yu. High Density 3D Integration, 2008 International Conference on Electronic Packaging Technology and High Density Packaging (ICEPT-HDP 2008), July 28–31, 2008.

[17] Katrin Persson, Arvid Hedvalls Backe, and Katarina Boustedt. Fundamental requirements on MEMS packaging and reliability. IEEE 8th International Symposium on Advanced Packaging Materials 2002: 1–7.

[18] Firebaugh, S.L., Charles, H.K., and Edwards, R.I. Packaging Considerations for Microelectromechanical Microwave Switches. IEEE Electronic Components and Technology Conference, 1 (2004): 862–868.

[19] Yufeng Jin, Jun Wei, Lim Peck Cheng, et al. Hermetic packaging of MEMS with thick electrodes by silicon-glass anodic bonding. International Journal of Computational Engineering Science, 4.2 (2003): 335–338.

[20] Runqing Yan and Yinghui Li. Basic theory of microwave technology (the second edition). Beijing: Beijing Institute of Technology Press, 1997.

[21] Matthew M. Radmanesh. Radio Frequency and Microwave Electronics Illustrated. Hong Kong: Pearson Education North Asia Limited, 2002.

[22] Liangjin Xue. Foundation for Millimeter Wave Engineering. Beijing: Defence Industry Publication, 1998.

[23] Osso, Rafael. Handbook of communications technology: the next decade, Boca Raton: CRC Press LLC, 2000.

[24] Theodore S. Rappaport. Wireless Communications—Principles and Practice. Englewood Cliffs, NJ: Prentice Hall, 1998.

[25] Lufei Ding and Fulu Geng. Radar Principle (second edition). Xi'an: XiDian University Press, 1995.

[26] Ferril Losee. RF Systems, Components, and Circuits Handbook. London: Artech House, 1997.

[27] Katehi, L.P.B., Rebeiz, G.M., Nguyen, C.T.-C. MEMS and Si-micromachined components for low-power, high-frequency communications systems. IEEE Transactions on Microwave Theory and Techniques, 1 (1998): 331–333.

[28] Nguyen, C.T.-C., Katehi, L.P.B., and Rebeiz, G.M. Micromachined Devices for Wireless Communications. Proceedings of the IEEE, 86.8 (1998): 1756–1767.

[29] R. J. Richards and H. J. De Los Santos. MEMS for RF/Microwave Wireless Application: The Next Wave. Microwave Journal, 44.3 (2001): 20–41.

[30] Cass, S. Large jobs for little devices [microelectromechanical systems]. IEEE Spectrum, 38.1 (2001): 72–73.

[31] Yiwen Liu, Zhijian Li, and Litian Liu. Application of MEMS Device in Communication Field. Electronic Science and Technology Review, 7 (1999): 23–27.

[32] Ehmke, J., Brank, J., Malczewski, A., et al. RF MEMS devices: a brave new world for RF technology. IEEE Emerging Technologies Symposium: Broadband, Wireless Internet Access, 2000: 1–4.

[33] J Héctor. Introduction to Microelectromechanical Microwave Systems. London: Artech House, 1999.

[34] Gabriel, M. Rebeiz. RF MEMS: Theory, Design, and Technology. New Jersey: John Wiley and Sons, 2003.

[35] William C. Tang. MEMS Programs at DARPA. Internet. Available: http://www.darpa.mil/MTO/MEMS, 2000.

[36] Chang Liu. Micro Electromechanical Systems (MEMS): Technology and Future Applications

in Circuits. In: 5th International Conference on Solid-State and Integrated Circuit Technology, 1998: 928–931.

[37] Rebeiz, G.M. and Muldavin, J.B. RF MEMS Switches and Switch Circuits. IEEE Microwave Magazine, 2.4 (2001): 59–71.

[38] Goldsmith, C., Randall, J., Eshelman, S., et al. Characteristics of micromachined switches at microwave frequencies. IEEE MTT-S International Microwave Symposium Digest, 2 (1996): 1141–1144.

[39] Goldsmith, C.L., Zhimin Yao, Eshelman, S., et al. Performance of low-loss RF MEMS capacitive switches. IEEE Microwave and Guided Wave Letters, 8.8 (1998): 269–271.

[40] Yao, Z.J., Chen, S., Eshelman, S., et al. Micromachined low-loss microwave switches. IEEE Journal of Microelectromechanical Systems, 8.2 (1999): 129–134.

[41] Pacheco, S., Nguyen, C.T., and Katehi, L.P.B. Micromechanical electrostatic K-band switches. IEEE MTT-S International Microwave Symposium Digest, 3 (1998): 1569–1572.

[42] Muldavin, J.B. and Rebeiz, G.M. Inline capacitive and DC-contact MEMS shunt switches. IEEE Microwave and Wireless Components Letters, 11.8 (2001): 334–336.

[43] Shyf-Chiang Shen, Caruth, D., and Feng, M. Broadband low actuation voltage RF MEM switches. In: 22nd Annual Gallium Arsenide Integrated Circuit (GaAs IC) Symposium, (2000): 161–164.

[44] Mihailovich, R.E., Kim M. Hacker, J.B. Sovero, et al. MEMs relay for reconfigurable RF circuits. IEEE Microwave and Wireless Components Letters, 11.2 (2001): 53–55.

[45] P.M. Zavracky, N.E. McGruer, and R.H. Morrison. Microswitches and microrelays with a view toward microwave applications. Journal of RF and Microwave Computer-Aided Engineering, (1999): 338–347.

[46] Bozler, C., Drangmeister, R. Duffy, S., et al. MEMS microswitch arrays for reconfigurable distributed microwavecomponents. IEEE MTT-S International Microwave Symposium Digest, 1 (2000): 153–156.

[47] Ye Wang, Zhihong Li, and McCormick, D.T. Low-voltage lateral-contact microrelays for RF applications. The Fifteenth IEEE International Conference on Micro Electro Mechanical Systems, (2002): 645–648.

[48] I.J. Bahl. Lumped elements for RF and microwave circuits. Boston, MA: Artech House, 2003.

[49] D.J. Young and B.E. Boser. A Micromachined Variable Capacitor for Monolithic Low-Noise VCOs. Solid-State Sensor and Actuator Workshop, (1996): 86–89.

[50] Dec, A. and Suyama, K. Micromachined Electro-Mechanically Tunable Capacitors and their Applications to RF ICs. IEEE Transactions on Microwave Theory and Techniques, 46.12 (1998): 2587–2595.

[51] R.L. Borwick, P.A. Stupar, and J. DeNatale. A high Q, large tuning range MEMS capacitor for RF filter systems. IEEE International Conference on Microelectromechanical Systems, (2002): 669–672.

[52] Jun-Bo Yoon and Nguyen, C.T.-C. A high-Q tunable micromechanical capacitor with movable dielectric for RF applications. International Electron Devices Meeting, (2000): 489–492.

[53] Jiang, H., Wang, Y., Yeh, J.-L.A., et al. Fabrication of high-performance on-chip suspended spiral inductors by micromachining and electroless copper plating. IEEE MTT-S International Microwave Symposium Digest, 1 (2000): 279–282.

[54] V.K. Varadan, K.J. Vinoy, and K.A. Jose. RF MEMS and Their Applications. London: John Wiley and Sons Ltd., 2003.

[55] Clark, J.R., Wan-Thai Hsu, and Nguyen, C.T.-C. High-Q VHF Micromechanical Contour-Mode Disk Resonators. International Electron Devices Meeting (2000): 493–496.

[56] Nguyen, C.T.-C. and Howe, R.T. CMOS micromechanical resonator oscillator. International Electron Devices Meeting (1993): 199–202.

[57] http://www.radantmems.com/radantmems/index.html, Oct 10th, 2007

[58] http://www.smalltimes.com, Oct 10th, 2007

[59] http: //www.darpa.mil, Oct 12th, 2007

[60] Wicht Technology Consulting. Technology Web Site. Internet. Available: www.wtc-consult.de, 2008.

[61] Yu Albert Wang. RF MEMS Switches and Phase Shifters for 3D MMIC Phased Array Antenna Systems. Ph.D. Dissertation, University of Cincinnati, 2002.

[62] D.G. Swanson and W.J.R. Hoefer. Microwave Circuit Modeling Using Electromagnetic Field Simulation. London: Artech House, 2003.

[63] Takahashi, K., Sangawa, U., and Fujita, S. Packaging using microelectromechanical technologies and planar components. IEEE Transactions on Microwave Theory and Techniques, 49.11 (2001): 2099–2104.

[64] Takahashi, K., Sangawa, U., and Fujita, S. Packaging using MEMS technologies and planar components. Asia Pacific Microwave Conference (2000): 904–907.

[64] Butler, J.T. and Bright, V.M. An embedded overlay concept for microsystems packaging. IEEE Transactions on Advanced Packaging, 23.4 (2000): 617–622.

[65] Gilleo, K. Overview of new packages, materials and processes. In: International Symposium on Advanced Packaging Materials: Processes, Properties and Interfaces (2001): 1–5.

[66] Drayton, Rhonda Franklin. The development and characterization of self-packages using micromachining techniques for high frequency circuit applications. Ph.D. dissertation, University of Michigan, 1995.

[67] T.R. Hsu. MEMS Packaging. London: INSPEC, 2004.

[68] R. Simons. Coplanar Waveguide Circuits, Components, and Systems. New York: John Wiley & Sons, 2001.

[69] Sung-Jin Kim, Young-Soo Kwon, and Hai-Young Lee. Silicon MEMS packages for coplanar MMICs. Asia-Pacific Microwave Conference (2000): 664–667.

[70] Immerman, M., Felton, L., Lacsamana, E., et al. Next generation low stress plastic cavity package for sensor applications. In: 7th Electronic Packaging Technology Conference, 1 (2005): 231–237.

[71] Gillot, C., Pornin, J.L., Arnaud, A., et al. Wafer level thin film encapsulation for MEMS. In: 7th Electronic Packaging Technology Conference, 1 (2005): 243–247.

[72] Linfa Liu. Next generation SiP of Renesas Technology. Electronics and Packaging, 5 (2005): 46.

CHAPTER 9

Reliability

9.1 Introduction

In general, reliability is the ability of a system or components and devices to perform its required functions under stated conditions for a specified period of time. Every microsystem product has its own specifications based on the product definition. It is supposed to be functional without failure for a given time when it is operated correctly in a specified environment. As we know, reliability requirements vary with products. An automotive controller is designed to last between 10 and 15 years, while only 3 to 5 years is expected for a mobile phone. In addition to these consumer products, a component for space application is required to work reliably over 20 or 30 years in the harsh environment of space radiation.

Microsystems have permeated every aspect of human life with the potential to bring everyone around the world into the digital age. We can hardly imagine a life without personal computers, mobile phones, automobiles, and so on. The definition of unreliability can be illustrated by a car thaty stops running. Failure of these consumer products may make some inconveniences or worse, e.g., the failure of an airbag may be fatal. Many critical areas, such as medical, aerospace, and military applications, have extremely high requirements for product reliability, where failure can have life-threatening consequences. The high cost of unreliability motivates an engineering study on microelectronics reliability.

Because "long-term" reliability is determined over a period of years, it is impossible to test these components for reliability for several years before they come into market. To ensure that an electronic component meets the reliability requirements, there are two existing approaches. [1,2] The first approach is design for reliability at the earlier stages of product design. This approach is based on the understanding of various potential failure mechanisms that could result in product failure; thus the product could be designed to minimize or eliminate the chances of failure by optimizing structures, materials, and processes. The second approach is to conduct an accelerated test on the finished product, which is a traditional industrial practice. The accelerated tests are commonly conducted within a reasonable amount of time under a well-controlled environment in a laboratory, which is effective for obtaining reliability data and for product qualification. If problems are found in reliability testing, the products must be rebuilt. It is a cost- and time-consuming approach compared with the first one.

Ideally, reliability testing should be conducted in the same environment as the product specified. However, in many cases, the designed lifetime is long enough to prohibit reliability testing under normal operating conditions. Even if time is not an issue, it is still impossible to simulate the actual working conditions in a laboratory. In accelerated tests, the devices are subjected to much higher "stress" than what they would experience under normal operating conditions. This accelerates the failure mechanisms, so that various failure modes can be observed within a reasonable period of time. Thus, reliability data can be collected within a much shorter time period. Different failure mechanisms dominate in different applications, and therefore it is not always necessary or possible to design against all failure mechanisms. Different electronic components are likely to experience different failure modes.

In the following sections, we will discuss the key issues of microsystem reliability in detail. It includes reliability-related theories, wafer/package related failure modes and mechanisms, reliability testing methods, and reliability standards.

9.2   Fundamentals of Reliability Methodology

Reliability design based on well-designed, well-understood, and thoroughly implemented accelerated tests, therefore, is a critical part of microsystems packaging. In general, reliability design addresses questions concerning when failure occurs, what causes failure, how to avoid failure, and what is the expected lifetime of an electronic device.

The failure rate is often used as a general index for representing semiconductor reliability. Semiconductor failure rates follow a bathtub curve as shown in Figure 9.1. The curve can be divided into three regions of initial failures, random failures, and wear-out failures according to the time of occurrence.

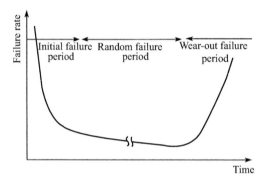

Figure 9.1   Semiconductor failure rates in a typical bathtub curve

1. Initial Failures

These are failures that occur within a relatively short time. Therefore, these failures are also called infant mortality failures, and this period is also called early failure period. They can be characterized by a decrease in the failure rate over time. The main causes of infant mortality failure are manufacturing or material defects. For example, a defect can be caused by tiny particles in a chip during the device production process, resulting in a device failure later. Because only devices with quality defects will fail early, removal of these failed devices will decrease the failure rate.

2. Random Failures

These are failures that occur at a fairly constant failure rate over a long period of time in between the early failure period and the wear-out failure period. After the devices with quality defects have been removed, the remaining high-quality devices operate stably. In this period, the failures can usually be attributed to randomly occurring excessive stress, such as power surges, other high-energy radioactive rays, etc.

3. Wear-out Failures

These are failures caused the aging of devices from wear and fatigue and occur due to the physical limitations of the materials that comprise semiconductor devices. The failure rate tends to increase rapidly in this period, and these failures are used to determine the device's lifetime. Therefore, the products are designed so that wear-out failures will not occur within their guaranteed lifetime.

Accordingly, it is important to reduce the initial failure rate to ensure long lifetime and durability against wear-out failures. Traditionally, individual companies implement quality controls and improvements as well as screening strategies, including electrical characteristics testing and burn-in tests. Furthermore, design-for-reliability is widely considered during the design and development stages, instead of the manufacturing stage alone.

9.2.1 Basic Reliability Theory[1,3]

The objective of any reliability study is to produce safe and reliable products. The evaluation and quantification of product reliability are prerequisites for reliability improvement and are necessary for determining trade-offs between reliability improvement and cost during design. Four measures are often used to quantitatively represent reliability. Their definitions are described below.

1. Cumulative failure distribution function (CDF): $F(t)$

CDF is the proportion of components, devices, parts, or elements that cease to perform their designed functions after being used for a period of time t. This can be expressed by the equation

$$F(t) = \frac{c(t)}{N},\tag{9.1}$$

where $c(t)$ is the number of failures to develop up to time t, and N is the total number of tested components.

2. Failure probability density function (PDF): $f(t)$

$f(t)$ denotes the probability of a device failing in the time interval dt at time t. From this definition PDF can be expressed by the equation

$$f(t) = \frac{dF(t)}{dt} = -\frac{dR(t)}{dt}.\tag{9.2}$$

As can be seen from the above equation, $R(t)$ and $F(t)$ can be calculated by taking the integral of the PDF, as below:

$$F(t) = \int_0^t f(t)dt,\tag{9.3}$$

$$R(t) = 1 - \int_0^t f(t)dt = \int_t^\infty f(t)dt.\tag{9.3a}$$

3. Reliability function: $R(t)$

The reliability function is the proportion of components (devices, parts, or elements) that continue to perform their designed functions and remain stable after time t. From the PDF definition, it is easy to get the reliability function as

$$R(t) = 1 - F(t).\tag{9.4}$$

The reliability function $R(t)$ is a monotonically decreasing function, and the $F(t)$ is a monotonically increasing function.

4. Failure rate: $\lambda(t)$

The failure rate $\lambda(t)$, also known as "hazard function," represents the probability of failure occurring in the next time interval for devices that have not yet failed when time t has passed. Using the concept of conditional probability, $P(B|A) = \frac{P(B \text{ and } A \text{ both occur})}{P(A)}$, we can derive that $\lambda(t)$ equals $f(t)/R(t)$ as shown below:

$$\lambda(t) = \frac{f(t)}{R(t)}.\tag{9.5}$$

The reliability function $R(t)$ can be expressed by $\lambda(t)$ as follows:

$$R(t) = \exp\left(-\int_0^t \lambda(t)\mathrm{d}t\right). \tag{9.6}$$

5. Mean time to failure (MTTF)

In general, once a component has failed, it cannot be repaired and used again. That is to say, it is a nonreparable component. MTTF is a basic measure of reliability for this type of device. It is the mean time expected until the first failure occurrence. MTTF is a statistical value and is meant to be the mean over a long period of time and a large number of units. It can be given by

$$\text{MTTF} = \int_0^\infty t f(t)dt \tag{9.7}$$

The overall relationship of $F(t), R(t), f(t), \lambda(t)$, and MTTF can be depicted as Figure 9.2.

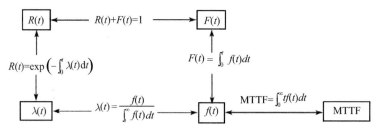

Figure 9.2   Overall relationships of basic reliability measurements

9.2.2   Distribution Used in Reliability Analysis

Since reliability stress tests are often destructive, only a sample population is used for reliability testing. As such, the assessment of reliability for the rest of the population is essentially statistical and probabilistic in nature. We will discuss commonly used distributions for semiconductor device reliability in detail.

1. Weibull Distribution

The Weibull distribution is a highly general-purpose distribution function that is expanded from the logarithmic distribution. In reliability data analysis, this model is frequently used to analyze life data in reliability tests, etc. The probability density function $f(t)$ and distribution function $F(t)$ of this distribution are as follows.

The three-parameter Weibull PDF is given by

$$f(t) = \frac{\beta}{\eta}\left(\frac{t-\gamma}{\eta}\right)^{\beta-1} \exp\left(-\frac{t-\gamma}{\eta}\right)^{\beta} \tag{9.8}$$

where $f(t) \geqslant 0, t \geqslant 0$ or $\gamma, \beta > 0, \eta > 0, -\infty < \gamma < \infty$, and $\eta =$ scale parameter, $\beta =$ shape parameter (or slope), $\gamma =$ location parameter.

We know through experience that no failures occur before a given test time γ, and after γ the total number of failed devices increases with time t (or more correctly, maintains a non-decreasing trend). The location parameter $\gamma = 0$ if we assume that the probability of failure is already above 0 immediately before testing. The three-parameter Weibull distribution will reduce to the two-parameter Weibull distribution, and its PDF is given by

$$f(t) = \frac{\beta}{\eta}\left(\frac{t}{\eta}\right)^{\beta-1} \exp\left(-\frac{t}{\eta}\right)^{\beta}. \tag{9.9}$$

Figure 9.3 shows the effect of different values of the shape parameter β on the shape of the PDF. One can see that the shape of the PDF can take on a variety of forms based on the value of β.

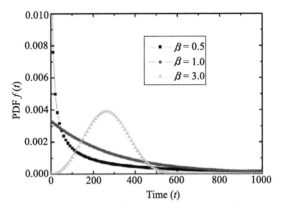

Figure 9.3 The effect of different values of β on the shape of the Weibull distribution's PDF

The equation for the three-parameter Weibull cumulative density function $F(t)$ is given by

$$F(t) = 1 - e^{-\left(\frac{t-\gamma}{\eta}\right)^{\beta}}. \tag{9.10}$$

The reliability function $R(t)$ for the three-parameter Weibull distribution is given by

$$R(t) = e^{-\left(\frac{t-\gamma}{\eta}\right)^{\beta}}. \tag{9.11}$$

The Weibull failure rate function $\lambda(t)$ is given by

$$\lambda(t) = \frac{f(t)}{R(t)} = \frac{\beta}{\eta}\left(\frac{t-\lambda}{\eta}\right)^{\beta}. \tag{9.12}$$

From Figure 9.4, we can see that the function form of the Weibull distribution is capable of representing different failure modes depending on the value of the parameter β. When $\beta = 1$, the PDF of the three-parameter Weibull reduces to that of the two-parameter exponential distribution as

$$f(t) = \frac{1}{\eta}e^{-\frac{t-\gamma}{\eta}}, \tag{9.13}$$

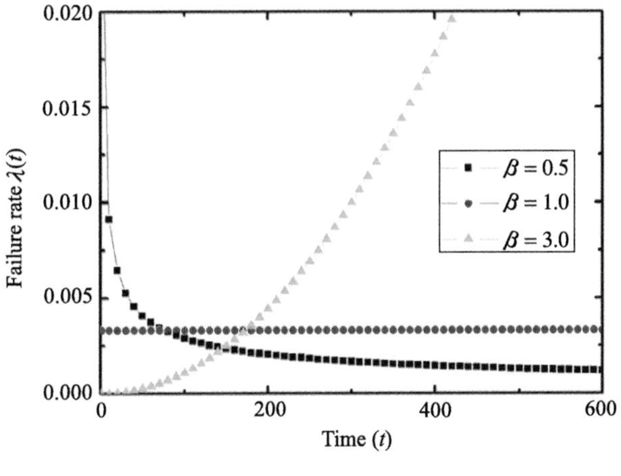

Figure 9.4 Effect of the shape parameter β on the failure rate of the Weibull distribution

where $\lambda = \dfrac{1}{\eta}$ is a constant.

When $\beta > 1$, the Weibull failure rate $\lambda(t)$ monotonically increases, representing a wear-out failure mode. For $1 < \beta < 2$, the $\lambda(t)$ curve is concave, consequently, the failure rate increases at a decreasing rate as time increases. For $\beta = 2$ there emerges a straight line relationship between $\lambda(t)$ and t, starting at a value of $\lambda(t) = 0$ at $t = \gamma$, and increasing thereafter with a slope of $\dfrac{2}{\eta^2}$. Consequently, the failure rate increases at a constant rate as time increases. Furthermore, if $\eta = 1$ the slope becomes equal to 2, and when $\gamma = 0$, $\lambda(t)$ becomes a straight line that passes through the origin with a slope of 2. Note that at $\beta = 2$, the Weibull distribution equations reduce to that of the Rayleigh distribution. When $\beta > 2$, the $\lambda(t)$ curve is convex, with its slope increasing as time increases. Consequently, the failure rate increases at an increasing rate as time increases, indicating wear-out life.

When $\beta < 1$, the Weibull failure rate $\lambda(t)$ is unbounded at $t = 0$ (or γ) and decreases thereafter monotonically and is convex, approaching the value of zero as $t \to \infty$ or $\lambda(\infty) = 0$. This behavior makes it suitable for representing the failure rate of units exhibiting initial failures, for which the failure rate decreases with increasing time. When encountering such behavior in a manufactured product, it may be indicative of problems in the production process, inadequate burn-in, substandard parts and components, or problems with packaging and shipping.

2. Normal Distribution

The normal distribution is commonly used for general reliability analysis and times-to-failure of simple electronic and mechanical components, equipment, or systems.

The PDF of the normal distribution is given by

$$f(t) = \frac{1}{\sigma\sqrt{2\pi}} e^{-\frac{1}{2}\left(\frac{t-\mu}{\sigma}\right)^2},$$

$$f(t) \geqslant 0, \ -\infty < t < \infty, \ \sigma > 0,$$

(9.14)

where μ is the mean of the normal times-to-failure and σ is the standard deviation of the times-to-failure.

The normal distribution is symmetrical about its mean value, and the standard deviation σ is the scale parameter of the PDF. As σ decreases, the PDF becomes narrower and taller. As σ increases, the PDF becomes broader and shallower, as shown in Figure 9.5. The

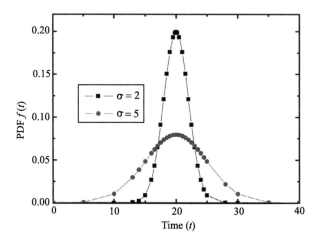

Figure 9.5 The typical normal distribution PDF with different standard deviation

probability of the value t falling within the range of $\pm\sigma$, $\pm 2\sigma$, $\pm 3\sigma$ to both side of μ is 68.26%, 95.44%, and 99.7%, respectively.

3. Logarithmic Normal Distribution

The logarithmic normal distribution is commonly used for general reliability analysis, cycles-to-failure in fatigue, material strengths, and loading variables in probabilistic design.

When the natural logarithms of the times-to-failure are normally distributed, then we say that the data follow the logarithmic normal distribution.

The PDF of the logarithmic normal distribution is given by

$$f(t) = \frac{1}{t\sigma'\sqrt{2\pi}}e^{-\frac{1}{2}\left(\frac{t'-\mu'}{\sigma'}\right)^2},$$
$$f(t) \geqslant 0, \ t > 0, \ \sigma' > 0, \ t' = \ln(t),$$

$$(9.15)$$

where μ' is the mean of the natural logarithms of the times-to-failure and σ' is the standard deviation of the natural logarithms of the times-to-failure.

4. Exponential Distribution

The exponential distribution represents the life distribution in the random failure region where the failure rate λ is a constant over time. The probability density function $f(t)$ and reliability $R(t)$ are as follows.

$$f(t) = \lambda e^{-\lambda t},$$
$$R(t) = e^{-\lambda t},$$

$$(9.16)$$

The value of the PDF function is always equal to the value of λ when t is 0.

9.3 Wafer and Packaging-related Failure Mode and Mechanisms

The key to continued reliability improvement is a fundamental understanding of the failure mechanisms that limit the lifetime of electronic products. For semiconductor electronic devices all failures are identified as electrical failures. However, the failure mechanisms can be thermal, mechanical, electrical, chemical, or a combination of these. Figure 9.6 presents the major failure mechanisms in microelectronic products, which shows microelectronic failures can be roughly classified into two categories.[1,4,5]

1. Overstress Failures

Failure due to stress exceeds the strength or the capacity of the component or material. From the mechanism point of view, mechanical overstress failures include brittle fracture, deformation, and delamination, while the electrical overstress failures are electrostatic discharge (ESD), electromagnetic interference (EMI), and radiation.

2. Wear-out Failures

These can originate during the whole application lifetime. Unlike overstress failures, they change gradually. Operation at lower stress over a specified period of time will result in cumulative damages that cause a component to lose its function eventually. They can be attributed to fatigue, creep, gate oxide breakdown, electromigration, corrosion, etc.

In general, the microelectronic chips are processed through the wafer process and packaging process. We can also consider major reliability issues from their processes. In the following sections, we will discuss wafer- and packaging-related failures and mechanisms separately.

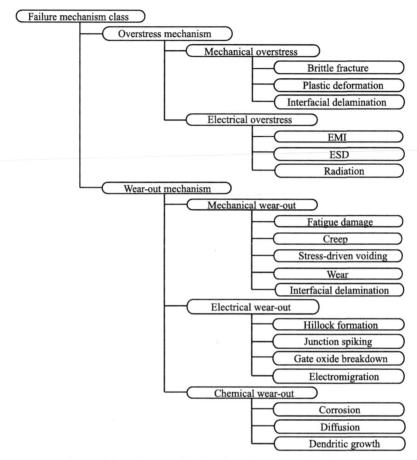

Figure 9.6   Failure mechanisms in microelectronic devices

9.3.1   Wafer-related Failures and Mechanisms[6]

1. Electrical Overstress (EOS)

Electrical overstress refers to the destruction of the device due to excessive voltage, current, or power. EOS is usually caused by improper application of excitation to the device, whether it is being tested or in the field. Improper excitation settings or voltage spikes in the excitation source are common causes of EOS damage. Latch-up and electrostatic discharge (ESD) are special cases of EOS.

2. ESD

Electrostatic discharge can occur when two objects at different potentials directly contact each other or come in close proximity. An ESD event can damage a device in many ways, such as conductor fusing, junction damage, dielectric/oxide ruptures, etc. There are three ESD models that are widely accepted today. They are the human body model (HBM), the charged device model (CDM), and the machine model (MM).

3. Junction Spiking

Junction spiking occurs when the Al migrates into the silicon substrate and has penetrated deep enough to short a positive-negative (p-n) junction. This phenomenon is due to the interdiffusion between two different interdiffusible materials in contact with each other, which are Al and Si in this case. Al migration can be reduced by adding Si to Al and forming a barrier metal between the Al and Si.

4. Electromigration (EM)

Electromigration refers to the gradual displacement of the metal atoms of a conductor as a result of the current flowing through that conductor. Because of the mass transportation of metal atoms from one point to another under the driving force of an "electron wind", this leads to the formation of voids at some points in the metal line and hillocks or extrusions at other points. Therefore it will result in failure due to an open circuit if the voids formed in the metal line become big enough to go through it, a short circuit if the extrusions become long enough to bridge the adjacent metal line, or simply by the change of resistance in the metal line. Electromigration is a function of current density. It is also accelerated by elevated temperature. Thus, electromigration is easily observed in Al metal lines that are subjected to high current densities at high temperature over time. Although electromigration has been widely recognized as an IC failure mode for many years, concern has increased over the probability of its occurrence toward the end of useful life. This concern coincides with the reduction of feature sizes into the submicron regime, with multiple levels of metallization. These technology trends result in increased interconnection current densities and device operating temperatures, both of which exacerbate electromigration. The complexity of assuring electromigration reliability is significant.[7−10]

5. Dielectric Breakdown

Dielectric breakdown refers to the destruction of a dielectric layer, usually as a result of excessive potential difference or voltage applied. It is usually manifested as a short or leakage at the point of breakdown. As we know, the EOS/ESD can expose the dielectric layer to high voltages that will cause the dielectric breakdown. Non-EOS/ESD-related dielectric breakdowns can be classified into either an early life dielectric breakdown (ELDB) or a time-dependent dielectric breakdown (TDDB), depending on when the dielectric breakdown occurs. ELDB usually occurs at the very beginning of operation, which is highly correlated with initial defects. However, TDDB failures are usually caused by the trap/charge generated in the gate oxide during high current/voltage stress. These gradually generated trap/charges will decrease the effective thickness of the gate oxide. At a critical point the oxide heats up and allows a greater current flow. This eventually results in an electrical and thermal runaway that quickly leads to the physical destruction of the gate oxide. As the device dimension scales rapidly, the gate oxide becomes thinner. Even though the supply voltages have decreased, the trend of miniaturization and improvement in performance results in higher electric fields across gate oxide. Therefore, gate oxide reliability becomes more important.[11−14]

6. Hot Carrier Injection

Either holes or electrons can become "hot" after they have gained very high kinetic energy from the strong electric field in areas of high field densities within microelectronic devices. Hot carriers can get injected and trapped in the gate oxide, resulting in device degradation. The effects of mechanical overstress (MOS) hot carrier degradation are an important issue, since device scaling has outpaced the scaling of the supply voltages resulting in increased electric fields in the channel and gate oxide. There are four commonly encountered hot carrier injection mechanisms. They are (1) the drain avalanche hot carrier injection; (2) the channel hot electron injection; (3) the substrate hot electron injection; and (4) the secondary generated hot electron injection.[15−17]

7. Soft Errors

Radiation is another failure mechanism that is especially important for dynamic random-access memories (DRAMs). The impurities of radioactive elements are present in the chip packaging material that emits α-particles with energies up to 8 MeV. The interaction of these α-particles with the semiconductor material results in the generation of electron-hole pairs. The generated electrons move through the device and are capable of wiping out the charge stored in a DRAM cell, causing its information to be lost. This is the major cause of soft errors in DRAMs. Current research has shown that high-density static random-access memories (SRAMs) also suffer from soft errors caused by α-particles.[18,19]

9.3.2   Packaging-related Failures and Mechanisms[20−23]

With the development of package technology, the various package types are emerging rapidly. Each technique has its own reliability issues. A common failure in wire-bonding plastic packages is delamination of the encapsulant molding component or die attachment, which leads to highly localized stress concentrations in wire bonds. Packaging-related failures include:

1. Die Cracking

Die cracking is the occurrence of fractures in or on any part of the die of a semiconductor device. Die cracks have a variety of causes, but they usually originate from die attach problems and/or mechanical stresses on the package that get transmitted to the die. Imperfections in the die itself can also result in cracks. Damage or defects at the backside of the die can serve as crack initiation points once the package is subjected to thermomechanical stresses. The backside grinding, wafer saw, and ejecting processes, will produce these microcracks, which significantly lower the fracture strength of the material.

2. Cratering

Cratering is a partial or total fracture of the silicon material underneath the bond pad. Cratering is commonly due to excessive stresses on the bond pads from an improper setup of wire-bonding parameters, such as temperature, ultrasonic power, bonding force, and bonding time. Thin metallization layers tend to exacerbate cratering. Temperatures of $250°C$ or above, lower ultrasonic power, and ramping the applied ultrasonic amplitude will help to minimize cratering.

3. Bond Fracture and Lift-off

Bond fracture and lift-off refer to any of several phenomena in which a wire bond becomes detached from its position, resulting in loss or degradation of electrical and mechanical connection between the bond pad and its bonding site. The weakest point in wire bonds is located at the heel of the crescent bond and at the neck of the ball bond. The imperfect bond pad surface is the primary cause of ball lifting, which includes contaminations, cratering, and silicon nodules on the surface of bond pads.

Interfacial delamination refers to the disbonding between a surface of the plastic package and that of another material. Interfacial delamination occurs as a result of poor adhesion between the molding compound and the die, the die passivation, the bond-wire, and lead-frame. The use of molding compounds with excessive mold release agent can also lead to delamination. Excessive mismatches between the thermal coefficient of the plastic and those of the lead-frame, die, and die attach material can also result in delamination. Residual stresses and thermomechanical stresses tend to aggravate the propagation of the interfacial microcracks and delamination. Moisture is also an acceleration factor on delamination. Delamination tends to result in highly localized stress concentrations and potential wire-bond fatigue fracture.

4. Package Cracking

Cracking can occur in many ways. For nonhermetic packages, the failure mode is largely a function of package-induced surface shear stresses, which results from a difference in CTE between the encapsulant, lead frame, and die. Package cracking occurs as a result of fatigue fracture of the encapsulant and due to a phenomenon called "popcorning." Factors that affect popcorn cracking tendency include the solder reflow temperature, the moisture content of the package, the dimensions of the die paddle, the thickness of the molding compound under the paddle, and the adhesion strength of the molding compound to the die and lead frame. The thermal mismatch between the encapsulant, lead frame, and die can produce large stresses to trigger the interfacial delamination. Moisture is often absorbed by the package from the environment during storage prior to the solder reflow. Intense pressure is generated by the vaporization of the internal moisture inside the package during solder reflow. The moisture ingress that collected by the delaminated surfaces will accelerate the delamination failure. The excessive stresses are relieved only after the fracture occurs, when the stress exceeds the encapsulant's characteristic fracture strength. For tape automated bonding, cracking results from the concentration of stress on the passivation layer under the TAB bump during bonding.

9.3.3   MEMS-related Failures and Mechanisms[24−27]

MEMS, as one of the most promising areas in the near future, is usually fabricated by IC compatible batch-processing technology. It is a technology combination of electronic circuit and micromachinery. Since the applications of MEMS include automotive, aerospace, biological/medical, and many other areas, its reliability is extremely important when it is used in critical applications. Compared to typical electronic devices, MEMS reliability covers both the electronic and mechanical parts, complicated by the interactions. The devices will be out of service when the moving parts fail to work, such as failed comb driver due to a broken finger, a particle of dust, or a metal particle, as shown in Figure 9.7. Since MEMS technology is still young, only a limited number of applications are successfully commercialized, such as accelerometers, pressure sensors, and ink-jet printer heads. Based on the most common generic elements used in MEMS devices, a list of common degradation/failure mechanisms of MEMS is given in Figure 9.8. MEMS reliability is a very young and quickly changing field. Currently there is not any dedicated equipment for MEMS reliability studies. Our knowledge of the physics of degradation and failure mechanisms is still limited.

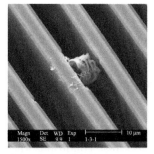

(a) broken finger   (b) A metal particle in a tweeter

Figure 9.7   Examples of failed comb drivers

- Fracture
  - Overload fracture
  - Fatigue fracture
- Creep
  - Applied stress
  - Intrinsic stress
  - Thermal stress
- Stiction
  - Capillary forces
  - Van der Waals molecular forces
  - Electrostatic forces
  - Solid bridging
- Electromigration
- Wear
  - Abrasion
  - Corrosion
- Degradation of dielectrics
  - Leakage
  - Charging
  - Breakdown
- Delamination
- Contamination
- Electrostatic discharge (ESD)

Figure 9.8   Common MEMS failure mechanisms

9.4   Reliability Qualification and Analysis

The qualification process for a new package is elaborate and includes a full set of reliability tests. The purpose of reliability qualification is primarily to ensure that the shipped products exhibit specified lifetime, functionality, and performance in the hands of the end users. Nevertheless, there are time and cost constraints. Because microelectronic devices require a long lifetime and low failure rate, testing under operation conditions would take unaffordable time and an excessively large sample size. Therefore, stresses beyond operation conditions are applied to devices to physically and/or chronologically accelerate causes of degradation. In this way, a device's lifetime and failure rates can be determined, and failure mechanisms can be analyzed. Such tests, referred as accelerated lifetime tests, are widely used to shorten the evaluation period and analyze mechanisms in detail. The stress test simulates in accelerated form, all conceivable loads such as initial ship-shock, storage, and use conditions to identify weaknesses in the package.

9.4.1   Accelerating Test[28−32]

It is necessary to note that failure mechanisms in accelerated tests differ somewhat from those that occur under actual usage conditions. In general, if the degradation mechanism is simple, acceleration is also simple, and lifetime and failure rates can be estimated relatively accurately. Complicated failure mechanisms, however, are difficult to simulate, even when the best efforts are made to accelerate stresses simultaneously. This is because different stress effects are interrelated. Therefore, analysis of acceleration data as well as estimation of lifetime and failure rates can be difficult. When performing accelerated lifetime tests, it is important to select test conditions that result in as few failure mechanism changes as possible, and that minimizes the number of failure mechanisms, making testing easy and simple. While the failure mechanism models provide a physical and mathematical representation of how the time-to-failure distribution will change by changing some factors, based on the accelerated test results and failure mechanism models, it is possible to project the lifetime under normal operating conditions. First linear regression is performed for the time-dependent cumulative failure rate using a certain distribution. And the lifetime is obtained from the time at which the reference cumulative failure rate is reached. The acceleration factor varies with the accelerated test conditions. Figure 9.9 illustrates this method on a Weibull probability plot.

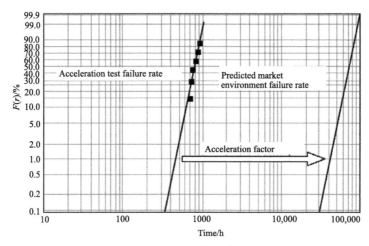

Figure 9.9   Failure rate projection based on Weibull distribution and accelerated test

9.4.2   Failure Mechanism Models[33−43]

As mentioned above, many failure mechanisms are thermal, mechanical, and electrical. Therefore, reliability tests must cover the types of stresses that the device will encounter in storage and operation, so as to provide meaningful lifetime data. And it is important to evaluate the acceleration factor of different stresses.

1. Temperature

Accelerated lifetime testing is closely associated with the physics of failure. Chemical kinetics is a basic chemical reaction model that describes the temperature dependence of failures. It is used with the Arrhenius model in accelerated lifetime testing of semiconductor devices in relation to temperature stress.

The Arrhenius model can be expressed as

$$\tau = A \exp\left(\frac{E_a}{kT}\right),\tag{9.17}$$

where E_a is activation energy, k is Boltzmann's constant, T is absolute temperature in Kelvin, and A is a constant. The acceleration factor can be found using the following equation:

$$AF = \exp\left[\frac{E_a}{k}\left(\frac{1}{T_{use}} - \frac{1}{T_{stress}}\right)\right].\tag{9.18}$$

From Arrhenius's equation it can be found that the failure rate is exponentially dependent on the temperature that the device is exposed to. Subjecting a component to a higher temperature in order to accelerate the aging process is called burn-in. Practical results have shown that a burn-in period of 50−150 hours at 125°C is effective in exposing 80%−90% of the component and production-induced defects (e.g., solder joints, component drift, weak components) and reducing the initial failure rate (infant mortality) by a factor of 2−10.

2. Voltage

Failure modes such as oxide film breakdown, hot carriers, Al corrosion due to humidity, and characteristic degradation due to mobile ions are accelerated by voltage. Of these, the failure mode that appears most dominantly as a result of voltage acceleration is oxide film breakdown. The voltage acceleration model for TDDB can be expressed using the following equation:

$$\tau = A \exp\left(-BE\right), \tag{9.19}$$

where E is the electric field applied to the gate oxide and A and B are constant parameters, The acceleration factor can be expressed as

$$AF = \exp\left[-B\left(E_{\text{use}} - E_{\text{stress}}\right)\right]. \tag{9.20}$$

3. Temperature Cycling

Microelectronic devices are made of various materials, and the coefficients of thermal expansion of these materials also vary widely. The difference between the coefficient of thermal expansion of each material will produce internal damage and accumulate each time when the device experiences a temperature cycling, which may lead to final failure. Temperature cycle tests, which apply a greater temperature difference than those normally experienced by the device, are effective as accelerated tests for evaluating damage caused by temperature differences. Tests for repeated thermal stress from the external environment or internal heat source include the temperature cycle test and the thermal shock test. In these tests, the device is subjected to repeated extreme temperature changes to determine temperature change resistance. The failure modes that occur during the tests include bonding opens, chip crack, etc. Lifetime related to such temperature cycling has been modeled by Coffin-Manson and can be express as follows:

$$\tau = A\left(\Delta T\right)^{m}, \tag{9.21}$$

where ΔT is the temperature difference and A and m are constants. The acceleration factor can be estimated by

$$AF = \left(\frac{\Delta T_{\text{stress}}}{\Delta T_{\text{use}}}\right)^{m}. \tag{9.22}$$

4. Humidity

Most microelectronic devices are encapsulated in plastic resin. The reliability of these devices largely depends on the humidity resistance of the package. Evaluation tests are roughly divided into two groups. The first method places the device in a humid atmosphere, and the second method applies bias to the device while exposing it to humidity or after moisture has penetrated it. The acceleration factor can be expressed as follows:

$$AF = \left(\frac{RH_{\text{stress}}}{RH_{\text{use}}}\right)^{m}, \tag{9.23}$$

where RH is the relative humidity and m is a constant. If temperature is added in the humidity test, both temperature and humidity acceleration factors should be considered. In the temperature/humidity/bias test, the voltage acceleration factor must be added. The junction heating effect can reduce the relative humidity. For example, a $5°C$ junction heating effect by bias can reduce the RH from 85% to 73%.

5. Current

Electromigration is the most well known current acceleration mode and is becoming more important as a failure mechanism as devices decrease in size and become highly integrated. The electromigration lifetime is generally expressed as follows:

$$\tau = AJ^{-n} \exp\left(\frac{E_a}{kT}\right), \tag{9.24}$$

where J is the current density and A and n are constants. Based on the above discussion, some commonly used acceleration models are summarized in Table 9.1.

Table 9.1 **Some commonly used package-related acceleration models**

Mechanism	Model	Assumptions
Temperature, humidity mechanisms	Peck's $TF = A_0 \cdot RH^{-N} \cdot \exp[E_a/kT]$ AF (ratio of TF values, use/stress) $AF = (RH_{stress}/RH_{use})^{-N} \cdot$ $\exp[(E_a/k)(1/T_{use} - 1/T_{stress})]$ When calculating AF, variables that remain constant from stress 1 to stress 2 will drop out of the equation.	AF = acceleration factor TF = time-to-failure A_0 = arbitrary scale factor V = bias voltage RH = relative humidity as % N = an experimentally determined constant E_a = activation energy for the mechanism (0.75 is conservative) k = Boltzmann's constant = 8.625×10^{-5} eV/ K T = temperature in Kelvin There are other models used for THB mechanisms that should be checked for fit to the data.
Thermal effects	Arrhenius $TF = A_0 \cdot \exp[E_a/kT]$ AF (ratio of TF values, use/stress) $AF = \exp[(E_a/k)(1/T_{use} - 1/T_{stress})]$	AF = acceleration factor TF = time-to-failure A_0 = arbitrary scale factor E_a = activation energy for the mechanism (0.75 is conservative) k = Boltzmann's constant = 8.625×10^{-5} eV/K T = temperature in Kelvin
Thermo-mechanical mechanisms	Coffin-Manson $N_f = C_0 \cdot (\Delta T)^{-n}$ AF (ratio of N_f values per stress cycle, stress/use) $AF = N_{use}/N_{stress} = (\Delta T_{stress}/\Delta T_{use})^n$	AF = acceleration factor N_f = number of cycles to failure C_0 = a material dependent constant ΔT = entire temperature cycle-range for the device, n = empirically determined constant, assumes that the stress and use ranges remain in the elastic regime for the materials. The Norris Landzberg modification to this model takes into account the stress test cycling rate.
Creep	$TF = B_0(T_0 - T)^{-n} \exp(E_a/kT)$ AF (ratio of TF values, use/stress) $= [(T_0 - T_{accel})/(T_0 - T_{use})]^{-n}$ $\exp[(E_a/k)(1/T_{accel} - 1/T_{use})]$	AF = acceleration factor TF = time-to-failure B_0 = process dependent constant T = temperature in Kelvin T_0 = Stress free temperature for metal ($\sim$metal deposition temperature for aluminum) n = 2−3, (n usually $\sim$5 if creep, thus implies $T < T_m/2$) E_a = activation energy = 0.5− 0.6 eV for grain-boundary diffusion, $\sim$1 eV for intragrain, k = Boltzmann's constant = 8.625×10^{-5} eV/K

9.4.3 Reliability-related Standards

There are numerous standards related to microelectronic device reliability. The most widely used standards are the military standards (Mil-Std) and EIA/JEDEC. Mil-Std is a collection of tests and evaluation procedures that are specially designed for reliability qualifications for military applications. EIA/JEDEC standards identify testing requirements that range from general to specific. For example, all plastic package testing must follow general guidelines specified by JESD 47, but only certain types of device packages may be required to undergo highly accelerated stress (HAST) testing. Figure 9.10 illustrates the hierarchy for EIA/JEDEC testing procedures and serves as the model for microelectronic products. The stress-test-driven qualification of integrated circuits (JEDEC Std. 47) guideline determines which tests a new design must undergo and helps product engineers identify and correct flaws that may arise. The guideline also identifies sample size requirements and testing qualifications for new devices to help expose process flaws.

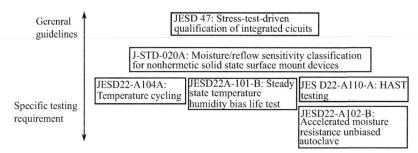

Figure 9.10   The hierarchy for EIA/JEDEC testing procedures

9.4.4   Reliability Test Methods

To ensure product reliability, each company establishes its own policy for reliability tests based on international standards. Evaluation of reliability begins in the planning stage. According to the customer's requirements, the appropriate levels of quality and reliability testing will be implemented into the product design and manufacturing stages and finally verified at the qualification stage. As we know, there are many industry-standard reliability tests already available.[44,53] The reliability test employed is chosen based on the failure mechanism of interest, as different stress tests accelerate different failure mechanisms. Nonetheless, most reliability tests use one or more of the following stress factors to accelerate failure: temperature, moisture or humidity, current, voltage, and pressure. In general, they are separated into electrical, thermal, and moisture-related tests that have been developed and refined over a period of time. In the following sections, the reliability test methods are discussed in detail.

1. Die/Process Reliability Tests

(1) High-temperature operating life test (HTOL).

According to JEDEC22-B-A108, the high-temperature operating life test is performed to accelerate failure mechanisms that are activated by temperature under bias. This test is used to predict long-term failure rates by temperature acceleration model. Prior to HTOL, all test samples are screened to standard electrical tests at low temperature and high temperature with prior burn-in. A typical stress voltage is 1.1 times normal operating voltage. Unless otherwise specified, the stress temperature is maintained at 125°C. Devices are tested at prescribed time points, such as 96 h, 168 h. The failure mechanism of HTOL results from the random failure/some early wear-out portions, mainly from defects of the chip surface, metallization, oxide, bulk, package, and ionic contaminated interconnect and wire bond.

(2) Infant mortality (IM).

Infant mortality testing determines the early failure rate of a specific product and process. The test conditions are basically the same as the high-temperature operating life test with an increased sample size to ensure an accurate failure rate. The test temperatures can be set between 125°C and 150°C, depending on product type and the test environment. The typical stress voltage is at least 1.2 times normal operating voltage. The failure rate data are used to determine a product burn-in strategy for each product and provide information for process improvement.

(3) Electrostatic discharge (ESD).

According to MIL-STD-883C, 3015; JEDEC22-A114, A115; and ESD-STM 5.3.1-1999, electrostatic discharge sensitivity (ESD) tests are designed to measure the sensitivity of each device with respect to electrostatic discharges that may occur during device handling. Various test methods have been devised to analyze ESD.

The human body model (HBM) is in accordance with the standard specified by MIL-STD-883C and JEDEC22-A114, while machine model (MM) is by MIL-STD-883C/JEDEC 22-A115 and the charge device model (CDM) is by ESD-STM 5.3.1-1999. The HBM is based on a high-voltage pulse (positive and negative) of longer duration, simulating discharge through human contact. The machine model is based on a high-current short duration pulse to simulate a device coming in contact with a charged surface. The charge device model is based on the phenomenon where the semiconductor device itself carries a charge or where the charge induced to the device from charged object near the device is discharged. It reproduces the discharge mechanism in the form closest to the discharge phenomenon occurring in the field.

(4) Soft error rate.

According to JEDEC Standard 89, semiconductor memory defects that can be recovered by rewriting the data are called soft errors. They result not only from the power supply line and ground line noise, but also from α-rays emitted from the trace amounts of uranium, thorium, and other radioactive substances contained in the package or wiring materials. There are two methods for evaluating soft errors: system tests, which consist of actually operating large number of samples, and accelerated tests by an external α-ray source. When evaluating the absolute soft error value it is necessary to conduct system tests. However, system tests require many samples and long time periods (typically, 1000 samples and 1000 hours or more). In contrast, accelerated tests allow evaluation in a short time, but have the problem that it is difficult to accurately obtain accelerated characteristics for an end-user environment.

2. Packaging Reliability Tests

(1) Highly accelerated stress test (HAST).

According to JEDEC22-B-A110, the highly accelerated stress test combines constant multiple stress conditions including temperature, humidity, pressure, and electrical bias. It is applied to quickly evaluate the reliability of nonhermetically packaged devices operating in humid environments. The HAST test is performed at 130°C and 85% relative humidity with applied bias. The electrical bias is used for powering the devices inside a HAST chamber. For HAST testing, Vcc is set to minimal conditions. The combination of multiple stress conditions accelerates the penetration of moisture through the package mold compound or along the interface between the external protective materials and the metallic conductors passing through the package. The presence of contaminants on the die surface such as chlorine greatly accelerates the corrosion process. Additionally, excessive phosphorus in the passivation will react under these conditions. The failure mechanism of HAST results from electrolytic corrosion of metal and contamination induced threshold shifts due to moisture penetration.

(2) Unbiased autoclave (Pressure cooker test [PCT]).

According to JEDEC22-B-A102, the autoclave test uses a temperature/humidity environment with no applied electrical bias to accelerate corrosion failure. It can be used to evaluate the moisture resistance of nonhermetically packaged units. Devices are subject to pressure, humidity, and elevated temperature to accelerate the penetration of moisture through the molding compound or along the interface of the device pins and molding compound. Expected failure mechanisms include mobile ionic contamination, leakage along the die surface, or metal corrosion caused by reactive agents present on the die surface. The autoclave test is performed in a chamber capable of maintaining temperature and pressure. Steam is introduced into the chamber until saturation, then the chamber is sealed and the temperature is elevated to 121°C, corresponding to a pressure of 33.3 psi. This condition is maintained for the duration of the test. Upon completion of the specified time, the devices are cooled, dried, and electrically tested. The failure mechanism of PCT results from the lost of hermeticity so that the metal line and bond pad get corroded due to passivation defects and ionic contamination (such as chlorine).

(3) Temperature cycling test (TCT).

According to JEDEC22-B-A104, the temperature cycling test is performed to determine the resistance of microelectronic devices to high- and low-temperature extremes. This stress test aims to simulate the extensive fluctuations in temperature to which devices and packages may be exposed. The changes in the temperature will cause damage between different components within the specific die and packaging system due to different thermal expansion coefficients. In the test, the devices are put inside a chamber where the temperature cycles between specified temperatures and held at each temperature for a minimum of 10 minutes. Temperature extremes depend on the conditions selected in the test method. The total stress corresponds to the number of cycles completed at the specified temperatures. The failure mechanism comes from the package cracking, die cracking, film cracking, bond wire lifting, die attachment problems and degradation of package hermeticity.

(4) Preconditioning test (moisture sensitivity).

According to JEDEC22-B-A112, A113, surface mount packages may be damaged during the solder reflow process when moisture in the package expands rapidly. Two test methods are used to determine which packages may be sensitive and what level of sensitivity exists. JEDEC test method A112 classifies devices into three groups: ① not sensitive, ② moisture sensitive, and ③ extremely moisture sensitive. JEDEC test method A113 verifies the reliability of devices exposed to a specified preconditioning process at various moisture levels by subjecting preconditioned devices to HAST, PCT, and TCT. After testing, the devices are assigned a moisture sensitivity level, which is useful in determining the proper storage and handling to ensure the reliability of the package process.

(5) Isothermal mechanical twisting (ITMT) test.[49]

An isothermal mechanical twisting (ITMT) test method was proposed to mimic similar solder joint fatigue failure mechanisms that occur in temperature cycling tests. In this method, a cyclic twisting load is imposed on an assembled printed circuit board (PCB) under an isothermal environment, as shown in Figure 9.11. The twisting deformation of the PCB is transferred to the solder joints and causes the solder joints to deform in a manner similar to that experienced in a temperature cycling test. Typically, the outmost solder joint on the diagonal will experience the highest strains.

It was shown that, with proper calibration on the twisting angle and testing temperature, this test method could be applied to assess the interconnection reliabilities of quad flat packs (QFP), ball-grid arrays (BGAs), and chip-scale packages (CSPs). Figure 9.12 shows a typical solder joint of a BGA assembly after the ITMT test. It shows that the location and feature of the crack is very similar to those that occur in the solder joints of a BGA assembly after a temperature cycling test.

Correlations of the results obtained from the ITMT tests with those obtained from the temperature cycling tests were conducted for various package types. It was found that the acceleration factor of a ITMT test against a temperature cycling test was typically 10−30 times in terms of cycles and around 100 times in terms of testing duration.

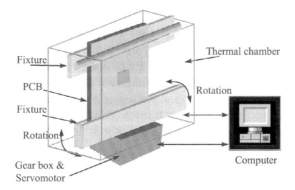

Figure 9.11   Schematic drawing of the ITMT testing system

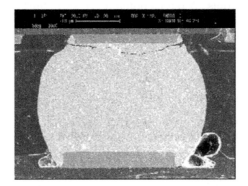

Figure 9.12   Solder joint failure after the ITMT test

(6) Vibration test.

In the automobile, military, and avionics industries, fatigue failure induced by vibration is one of the concerns. In order to understand the vibration reliability characteristics of PBGA assemblies, a detailed investigation on the testing and analysis techniques was performed.[49] At first, a bare PCB and a PCB assembled with plastic BGA (PBGA) modules were tested and analyzed separately to determine the influence of the PBGA modules on the PCB's dynamic properties. It was found that assembling PBGA modules onto the PCB increased the modal frequencies, but the modal shape vectors remained unchanged. This was mainly due to the increase in stiffness rather than the total mass.

In the vibration test, the specimen was clamped on two opposite sides using rectangular steel blocks and mounted on a solid test fixture, which was bolted to a vibration shaker, as shown in Figure 9.13.

Figure 9.13   Vibration test of a PBGA assembly

The vibration reliability test of the PBGA assemblies indicated that failure usually occurred at the corner joints. The die size had little effect on the vibration fatigue life, which is different from that in a temperature cycling test.

Another important observation is that the PCB assembly shows strong nonlinearity of dynamic response, which conflicts with the assumption commonly used by most researchers that the dynamic response of PCB assemblies has a linear relationship with the external vibration excitation. It indicates that such an assembly must be considered a nonlinear vibration system in its reliability assessment by analytic or numerical methods.

9.4.5   Case Study I: Solder Joint Design Methodology[49]

The development of solder joint reliability design methodologies can provide design guidelines before any prototype development starts. This requires an understanding on solder joint formation, constitutive models for solder materials, fatigue life prediction models, and simulation and verification tools. A public domain software, Surface Evolver, has been used in the prediction of the solder joint geometry of QFP, BGA, and flip chip. This software can be linked with finite element method (FEM) packages, such as Marc and Abaqus, for the simulation of stress and strain in a critical solder joint. The results are then used to predict solder joint reliability.

1. Materials Constitutive Models

One of the major problems faced in the early 1990s for the reliability assessment of solder joints was absence of reliable material data for FEM simulation. Different values, for example, of Young's modulus, at the same testing temperature were reported by different researchers. The difference was found to be caused by the different strain rates used in their tests through a comprehensive testing program. Emprical formulae were derived to represent the effect of temperature and strain rate on the mechanical properties, such as Young's modulus E, Yield stress σ_y, and ultimate tensile strength σ_{uts}, of the eutectic solder, 63/37 Sn/Pb. These formulae are given as follows:

$$E = (-0.006T_c + 4.72)\log \dot{\varepsilon} + (-0.117T_c + 37), \tag{9.25}$$

$$\sigma_y = \begin{cases} (-0.22T_c + 62)[\dot{\varepsilon}]^{(8.27 \times 10^{-5}T_c + 0.0726)} & \dot{\varepsilon} \geqslant 5 \times 10^{-4} \\ (0.723T_c + 105.22)[\dot{\varepsilon}]^{(1.624 \times 10^{-3}T_c + 0.1304)} & \dot{\varepsilon} < 5 \times 10^{-4} \end{cases}, \tag{9.26}$$

$$\sigma_{uts} = \begin{cases} (-0.25T_c + 76)[\dot{\varepsilon}]^{(2.06 \times 10^{-4}T_c + 0.0825)} & \dot{\varepsilon} \geqslant 5 \times 10^{-4} \\ (0.91T_c + 128.775)[\dot{\varepsilon}]^{(1.581 \times 10^{-3}T_c + 0.1337)} & \dot{\varepsilon} < 5 \times 10^{-4} \end{cases}, \tag{9.27}$$

where T_c is the Celsius temperature in centigrade and $\dot{\varepsilon}$ is the strain rate.

Existing creep models were found unable to predict the creep deformation of solders accurately over the temperature and stress ranges for typical testing and service conditions of electronics products. This is because several creep mechanisms may occur at the same time. These mechanisms operate independently, and the fastest process dominates. A deformation mechanism map for 63/37 Sn/Pb solder alloy is shown in Figure 9.14.

Unified constitutive models were proposed to describe the creep behavior of the eutectic solder alloy. For dislocation controlled creep deformation, the shear creep strain rate for 63/37 Sn/Pb is given by

$$\frac{d\gamma}{dt} = C_1 \frac{G}{T}\left[\sinh\left(\alpha \frac{\tau}{G}\right)\right]^{n_l} \exp\left(-\frac{Q_l}{RT}\right) + C_h \frac{G}{T}\left[\sinh\left(\alpha \frac{\tau}{G}\right)\right]^{n_h} \exp\left(-\frac{Q_h}{RT}\right), \tag{9.28}$$

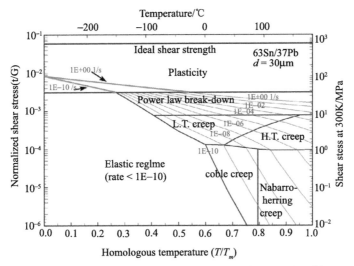

Figure 9.14   Deformation mechanism map of the eutectic solder

where Q_l and Q_h are the activation energies, and the subscripts l and h represent low-temperature dislocation glide process controlled creep and high-temperature dislocation climb process controlled creep. The constants in Equation 9.4 were determined from the experimental results as $C_1 = 2 \times 10^{-5}$, $C_2 = 0.25, \alpha = 1289, n_l = 5, n_h = 3, Q_l = 48.5$ kJ/mol, $Q_h = 81.5$ kJ/mol, $G = (24782 - 39.63T) \times 10^6$ is the temperature dependent shear modulus, and T is the absolute temperature in Kelvin. For diffusion controlled creep deformation at very low stress level, the shear creep strain rate is given by

$$\dot{\gamma} = A_1 \frac{G}{T} \frac{\tau}{G} \exp\left(-\frac{Q_b}{RT}\right) + A_2 \frac{G}{T} \frac{\tau}{G} \exp\left(-\frac{Q_m}{RT}\right), \tag{9.29}$$

where $A_1 = 1 \times 10^{-17}, A_2 = 1 \times 10^{-8}, Q_b = 54.5$ kJ/mol the activation energy for grain boundary diffusion, and $Q_m = 87.5$ kJ/mol the activation energy for matrix diffusion. It was found that Equation 9.28 and 9.29 can represent very well published experimental testing data.

Equation 9.28 and 9.29 are further combined into one unified model to cover the whole stress range and applied succcessfully to the fatigue life prediction of solder joints of PBGA assemblies under a temperature cycling test.

2. Fatigue Life Prediction Model

Most of the existing fatigue life prediction models for eutectic solder were based on the isothermal test results. The material constants in the models were usually averaged over the testing temperature. In some models, a frequency term was introduced to account for the effect of cyclic frequency. It was found that the so-called constants in these models varied greatly with either temperature or frequency.

In order to develop a fatigue life prediction model independent of testing temperature and frequency, the fatigue tests of eutectic solder specimens were carried out at different frequencies from 10^{-4} Hz to 1 Hz and at different temperatures from $-40°C$ to $150°C$ with a total strain controlled at values of 0.5% to 50%. Through analysis of the test results using existing fatigue models, a new model was proposed:

$$\left[N_f \nu^{(k-1)}\right]^m \frac{W_p}{2\sigma_f} = C, \tag{9.30}$$

where W_p is the nonlinear strain energy density, ν is the frequency, σ_f is the flow stress, m is the fatigue exponent, k is the frequency exponent, and C is the ductility coefficient. It was found that this model could describe the fatigue behavior of tin-lead eutectic solder very well over the test temperatures and frequencies. Moreover, the constants m, k, and C no longer varied with the temperature or frequency. Application of the constitutive models and the fatigue life model to PBGA solder joints provided a good estimation of the fatigue life in temperature cycling tests.

3. FEM Simulation

Finite element techniques have been used extensively in our work for process capability study, solder joint design, virtual design of experiments (DOE) for package design and materials selection, and for the reliability assessment of various packages in temperature cycling tests, vibration tests, and ITMT tests. It was shown that ignoring the strain rate effect in simulation could result in hundreds of times difference in the predicted fatigue life. This highlights the importance of the consideration of the strain rate effect in the constitutive models.

The effect of mesh density has been studied. It is known that due to the nonlinearity of the material and the stress singularity at the corner of a joint, the denser the mesh around the corner where the highest strain is, the higher the strain and strain energy density. This would result in a large difference in the predicted life. On the other hand, the volume averaged creep strain energy density shows much smaller variation for different mesh densities and it converges quickly to a value with the increase of mesh density. Therefore, a life prediction model based on the volume-weighted nonlinear strain energy density provides a life prediction that is less sensitive to the mesh density.

4. Deformation Measurement of IC Packages as a Tool of Reliability and Failure Analysis

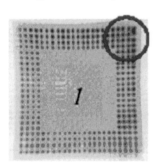

Figure 9.15   Solder bridging failures of BGA during reflow process

Owing to the complicated thermomechanical behaviors of packaging materials and the structure of IC packages, various simplifications are usually made in FEM simulations; as a result, people can not accept the simulation results with confidence. To verify the simulation accuracy or efficiency, deformations of IC packages need to be tested and compared with those from simulations. On the other hand, deformations of IC packages are usually needed for failure analysis, for example, solder bridging and leaking usually occur during the solder reflow process as shown in Figure 9.15. Engineers tend to attribute these failures to overlarge warpage of IC packages. In situ warpage measurement becomes a tool of failure analysis under this situation. Optical measurement techniques are developed for such applications, such as moiré interferometry, [50] Twymen/Green interferometry, [51] the Shadow moiré method, [52] etc. In this section, moiré interferometry and its applications are introduced. The advantages of optical measurements include their whole-field and noncontact testing.

A schematic diagram of the moiré interferometry system is shown in Figure 9.16.[1] In this configuration, a high-frequency cross-line diffraction grating is replicated on the surface of the specimen and it deforms together with the underlying specimen. Coherent beams B_1 and B_2 lie in the horizontal plane incident on the specimen grating at a specific angle and create a virtual reference grating in their zone of intersection. The deformed specimen grating and the reference grating interact to produce the moiré fringe pattern (U field pattern). Beams

B_3 and B_4 in the vertical plane create anther set of reference gratings, which interacts with the second set of lines in the 2-D specimen grating and produces another set of fringe patterns (V field pattern). The fringe patterns represent contours of constant U and V displacements in the orthogonal x and y directions, respectively. The displacement can be determined by

$$U = \frac{1}{f}N_x, \quad V = \frac{1}{f}N_y, \tag{9.31}$$

where f is the frequency of the virtual reference grating, and N_x, N_y are fringe orders in the U and V field moiré fringe patterns, respectively. Usually a virtual grating with a frequency of 2400 lines/mm is used, which provides a basic resolution of 0.417 μm/fringe.

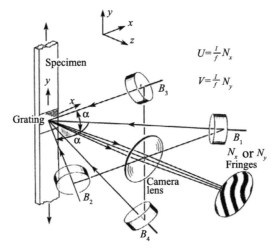

Figure 9.16   Schematic diagram of the moiré interferometry system

With the moiré interferometry system, the solder joint deformation under thermal cycling or a combination of thermal/humidity loading or thermal/mechnical loading can be measured. As an example, the solder joint deformations of a PBGA assembly under the temperature cycling test from $-40°C$ to $125°C$ were measured at different temperatures. The temperature profile in the measurement was similar to that used in an actual temperature cycling test, as shown in Figure 9.17. Letters, A, B, C, $\ldots$, P indicate the temperature levels at which the measurements were made.

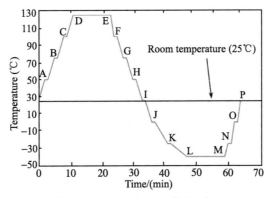

Figure 9.17   Temperature cycling profile in the measurement

A typical V field moiré fringe pattern obtained at $100°$C for the solder joint underneath the corner of the die is shown in Figure 9.18, which is the weakest joint in the assembly. Similarly, the U field moiré fringe pattern at the same temperature could be obtained. Based on both U field and V field fringe patterns, the shear strain can be calculated. Repeating the process at different temperature levels, the variation of the total shear strain with temperature can be determined. A typical result is shown in Figure 9.19, where the shear strains of the weakest joint were measured at different points of the first four cycles.

Figure 9.18   A typical moiré pattern at $100°$C of the weakest joint

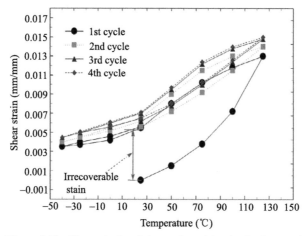

Figure 9.19   Shear strain of the weakest joint in the four cycles

By applying the life prediction model described above, the fatigue life of the solder joint was predicted to be 4699 cycles, which is close to the mean time to failure (MTTF) of the same packages, 4455 cycles, in the actual temperature cycling test.

9.4.6   Case Study II[53]

The microcomb structure made by bulk-Si undertakes a standard sample in MEMS or MOEMS, since it takes great applications in MEMS as a senor or an actuator normally. Sensors with comb structures like resonators, accelerometers, and gyros can be easily produced by standard bulk- Si processes and realize a 2D or 3D direction sensing with a high level sensitivity, high reliability, and an expandable ability. The comb actuator also has the most popular application in RF MEMS for its integratability with IC in a die. Therefore, this case will discuss the reliability issue by using MEMS devices with comb structures as samples.

1. Roadmap

The roadmap of how to study the complex reliability problems in MEMS center of Peking University follows four parts as shown in Figure 9.20; FA, test structure, lifetime, and harsh environment. FA involves investigating failure modes, failure analysis, and building a failure mechanism. These will be performed in three levels in turns of die level, module level, and system level. The test structure part relates the studies of designing test structure like stress gauge to extract the parameters of the materials, processes, etc. These studies will be carried out to establish a database of the parameters. The lifetime is the studies about mean time to failure (MTTF), Mean time before failure (MTBF), storage lifetime, accelerated stress test, modeling, etc. The harsh environment will be explored to find some novel method to relieve stresses and promote the quality of products. As the most important one and the most urgent one, the FA of MEMS was studied first.

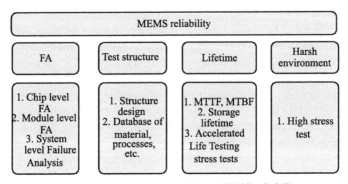

Figure 9.20   Subjects in studying MEMS reliability

2. FA

FA is a procedure involving failure sites localization, root cause investigation, and failure model establishment. As the first step of FA, the failure sites localization was pursued by using a series of techniques in three levels of production: die level, module level, and system level, as shown in Figure 9.21. While many FA techniques are based on IC techniques, some of the MEMS FA techniques are quite different from the ones of IC, since FA of MEMS involves not only electrical analysis, but also mechanical, material, and circumstance analysis. In this section, the techniques for MEMS failure analysis were discussed in three parts: (1) Failure localization techniques; (2) Characterization techniques; (3) Subsidiary techniques. In addition, some FA cases using parts of these techniques were presented in Figure 9.21 as (1) Yield rate investigation; (2) Failures localization in die level; (3) Failure analysis in die level; (4) Failure investigation in module level; (5) Exploratory experiment of Particle Impact Noise Detection test.

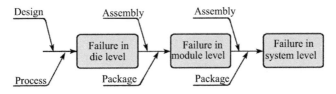

Figure 9.21   Failure localization in production

In MEMS failure studies, the main FA techniques can be divided into failure modes identification, failure characterization, and other subsidiary techniques for preparing samples.

3. Yield Rate Investigation

In this study, the failures were identified in macroscopic analysis and microscopic analysis first. The techniques for locating failure used in MEMS FA, such as optical microscopy and scanning electron microscopy (SEM), are both popular in MEMS FA to detect, surface failures, and the confocal IR image, X-ray inspection, in circuit tester (ICT) and SAM are used in locating the defects under the surface layer.

In the following steps, the spotted failures were determined to be certain kinds of failure modes, and the root causes were found by characterization techniques, where AFM, Raman spectroscopy, X-ray diffraction, Nano indenter, optical scanning profilometer, and full field optical profilometer can be used in detecting the deformation of the structure to calculate the residual stress, and material characterization techniques like Transmission electron microscope, Energy dispersive X-ray spectroscopy, Electron Energy Loss Spectroscopy), Auger Electron Spectroscopy., and Secondary ion mass spectrometry can be used to determine the contaminations or other special failures related to the material.

In large-scale production, failures of the comb structure were investigated to improve the reliability. The static capacitance test and the surface check by microscope and SEM were carried out to screen the unqualified structures. In the screening, almost all of the failure modes were found to be: overly thin finger, hogging finger, structure disjunction, etch pits, and some intact structures. The former three failures turned out to be over etched, and the broken fingers normally were induced by random shock operating or man-made errors. However, about 19% structures in the die level after dry etching release showed a static capacitance that was too large, as shown in Figure 9.22. Since the structures looked like qualified ones with a perfect surface, the failures needed to be investigated and analyzed in detail.

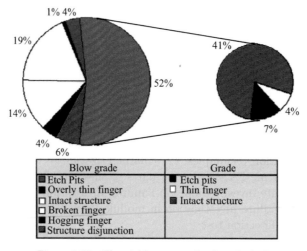

Figure 9.22   The yield map-graph of comb devices

4. Failure Idetification

Through surface inspection, failures like contaminations, etch pits, and failures by over etching were recognized. Figure 9.23 showed some contaminations by SEM/EDX in 10 kV, where (a) was a SEM photo of a contamination and (a′), an EDX curve with a Si peak, shows that the contamination was a broken comb. Similarly for (b)−(e): (b′) with C peak shows us that the contamination is a kind of organics material. The organic matter turned out to be photosensitive resist (PR); (c′) with Si and O peaks shows it could be some dust of

SiO$_2$, etc. (d$'$) with peaks of C, Na, Cl, etc. shows that the contamination must be a furfur from operator; (e$'$) with peaks of Fe, Ni, and Cr shows it is a metal particle that could come from stainless tweezers.

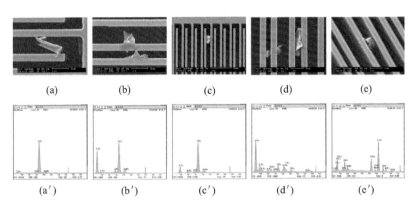

(a)   (b)   (c)   (d)   (e)

(a$'$)   (b$'$)   (c$'$)   (d$'$)   (e$'$)

Figure 9.23   SEM photos of contaminations and material analysis by EDX. (a) and (a$'$) SEM photo and EDX analysis of a broken comb, respectively. (b) and (b$'$) Some PR points. (c) and (c$'$) A particle of dust. (d) and (d$'$) A furfur from operator. (e) and (e$'$) A metal particle from tweezers

9.5   Summary

The development of the microelectronics industry has been accurately predicted by the famous Moore' Law.[53] Many technologies were developed to meet customers' demands, not only in terms of performance, but also reliability. Because the devices are becoming more miniaturized with high density, there are many new materials, novel structures, and processes to be applied; the interaction among them will be much more complicated, [54] some of the basic theories may need to be modified to deal with the new applications, so that ensuring that their reliability has been considered properly. Understanding the newly emerging failure modes and degradation mechanisms is very challenging and rewarding. In addition, to reduce costs and speed up the time-to-market and improve reliability performance, the next generation reliability technology must be able to achieve substantial cost reduction in design, reliability analysis and test, and product qualifications. Various reliability techniques, such as virtual projection, simulation, and accelerated test, should be integrated into the every stage through the whole product development process.

Questions

(1) What is the bathtub curve? How can you use this curve to predict failure problems in microsystems applications?

(2) Please describe the main procedure in studying MEMS reliability?

(3) How many reliability test methods have been widely used in microsystems engineering? Please summarize their applications in reliability study.

References

[1] Ramesham, R., Ghaffarian, R., and Ayazi, F., Fundamentals of Microelectromechanical Systems, Chapters 5 and 22. New York: McGraw-Hill, 2001.

[2] J. Prendergast, E. Murphy, M. Stephenson. Building In Reliability—Implementation and Benefits. International Journal of Quality and Reliability Management, 13(5) (1995): 77–90.

[3] Y. Gu. Math for Reliability Engineering. Beijing: Publishing House of Electronics Industry, 2004.

[4] R. Plieninger, M. Dittes, and K. Pressel, Modern IC Packaging Trends and Their Reliability Inplications. Microelectronics Reliability, 46 (2006): 1868–1873.

[5] H. Reichl, A. Schubert, and M. Topper. Reliability of Flip-Chip and Chip Size Package. Microelectronics Reliability, 40 (2000), 1243–1254.

[6] B. Wunderle and B. Michel. Progress in Reliability Research in the Micro and Nano Region. Microelectronics Reliability, 2006, 46: 1685–1694.

[7] J. R. Black. Physics of Electromigration. IEEE International Reliability Physics Symposium, 1974: 142–149.

[8] F. M. DeHeurle. Electromigration and Failure Electronics—An Introduction. IEEE International Reliability Physics Symposium, 1971: 109–118.

[9] P. B. Ghate. Electromigration Induced Failures in VLSI Interconnects. IEEE International Reliability Physics Symposium, 1982: 292–299.

[10] J. R. Black. Electromigration failure models in aluminum metallization for semiconductor devices. Proceedings of the IEEE, 57(9): 1587–1969.

[11] J. H. Chen, C.T. Wei, S.M. Hung, S.C. Wong, and Y.H. Wang. Breakdown and stress-induced oxide degradation mechanisms in MOSFETs. Solid State Electronics, 2002, 46: 1965–1974.

[12] D.J. Dumin, J.R. Maddux, R.S. Scott, and R. Subramoniam. Model relating wearout to breakdown in thin oxides. IEEE Transactions on Electron Devices, 1994, 41: 1570–1580.

[13] H. Iwai and S. Ohmi. Trend of CMOS downsizing and its reliability. Microelectronics Reliability, 2002, 42: 1251–1258.

[14] I.C. Chen, S. E. Holland, and C. Hu. Electrical Breakdown of Thin Gate and Tunnelling Oxides. IEEE Transactions. on Electron Devices, 1985, 32(2), 413–421.

[15] C. Hu, et al. Hot-electron-induced MOSFET degradation—Model, monitor and improvement. IEEE Transactions on Electron Devices, 1985, 32: 375–385.

[16] A. Acovic, G. La Rosa, and Y.C. Sun, A review of hot-carrier degradation mechanisms in MOSFETs. Microelectronics Reliability, 1996, 36: 845–869.

[17] H. Aono, E. Murakami, et al. NBT induced Hot Carrier (HC) Effect: Positive Feedback Mechanism in p-MOS FET's Degradation. IEEE International Reliability Physics Symposium, 2002, p. 79–85.

[18] T. C. May, et al. A New Physical Mechanism for Soft Errors in Dynamic RAMs, IEEE International Reliability Physics Symposium, 1978, p. 33–40.

[19] S. Yamamoto et al. Neutron-Induced Soft error in Logic Devices Using Quasi-Mono energetic Neutron Beam. IEEE International Reliability Physics Symposium, 2004, p. 305–309

[20] J. L. Flood. Reliability Aspects of Plastic Encapsulated Integrated Circuits. IEEE International Reliability Physics Symposium, 1972, p. 95–99.

[21] JEDEC standard 29 Failure Mechanism Driven Reliability Monitoring of Silicon Devices.

[22] A.S. Chen, L.T. Nguyen, and S.A. Gee. Effect of material interactions during thermal shock testing on IC package reliability. Electronic Components and Technology Conference, 1993: 693–700.

[23] J.C. Yang, C.W. Leong, J.S. Goh, and C.K. Yew. Effects of molding compound properties on lead-on-chip package reliability during IR reflow. Electronic Components and Technology Conference, 1996: 48–55.

[24] D. Tanner, et al. MEMS Reliability in Shock Environments. IEEE International Reliability Physics Symposium, 2000, 129–138.

[25] R. Muller-Fiedler, U. Wagner, and W. Bernhard. Reliability of MEMS—a methodical approach. Microelectronics Reliability, 2002, 42: 1771–1776.

[26] W. Merlijn van Spengen. MEMS reliability from a failure mechanisms perspective. Microelectronics Reliability, 2003, 43: 1049–1060.

[27] Roland Muller-Fiedler and Volker Knoblauch. Reliability aspects of microsensors and micromechatronic actuators for automotive applications. Microelectronics Reliability, 2003, 43: 1085–1097.

[28] N. Lycoudes. Practical Uses of Accelerated Testing at Motorola. IEEE International Reliability Physics Symposium, 1975, p. 257–259.

[29] B. Foucher, J. Boullie, B. Meslet, and D. Das. A review of reliability prediction methods for electronic devices. Microelectronics Reliability, 2002, 42: 1155–1162.

[30] J. Gunn and S. Malik. Highly Accelerated Temperature and Humidity Stress Test Technique (HAST). IEEE International Reliability Physics Symposium, 1981: 45–81.

[31] D. Danielson, et al. HAST Applications: Acceleration Factors and Results for VLSI Components. IEEE International Reliability Physics Symposium, 1989: 144–121.

[32] J. W. McPherson and D. A. Baglee. Accelerated Factors for Thin Gate Oxide Stressing. IEEE International Reliability Physics Symposium, 1985: 1–8.

[33] J. C. Lee, I.C. Chen, and C. Hu. Statistical Modeling of Silicon Dioxide Reliability. IEEE International Reliability Physics Symposium, 1988: 131–138.

[34] R. Moazzami, J. C. Lee, and C. Hu. Temperature Acceleration of Time Dependent Dielectric Breakdown. IEEE Transactions. on Electron Devices, 1989, 36(11): 2462–2465.

[35] N. Shino and M. Isumi. A Lifetime Prediction Model Using Series Model and Acceleration Factors for TDDB Failures of Thin Gate Oxides. IEEE International Reliability Physics Symposium, 1993: 1–6.

[36] H. Satake, et al. Impact of TDDB Distribution Function on Lifetime Estimation in Ultra-Thin Gate Oxides, Solid State Devices and Materials, 2000, p. 248–249.

[37] J. Prendergast, E. Murphy, M. Stephenson. Predicting Oxide Reliability From In Line Process Statistical Reliability Control. IEEE International Reliability Workshop, 1996: 42–49.

[38] F. N. Sinnadurai. The Accelerated Aging of Plastic Encapsulated Semiconductor Devices in Environment Containing A High Vapor Pressure of Water. Microelectronics and Reliability, 1974, 12: 1–5.

[39] D. S. Peck and C. H. Zierdt. Temperature-Humidity Acceleration of Metal-Electrolysis in Semiconductor Devices. IEEE International Reliability Physics Symposium, 1973, p. 146–152.

[40] S. Peck. Comprehensive Model for Humidity Testing Correlation. IEEE International Reliability Physics Symposium, 1986: 44–50.

[41] J. Ferro. An Accelerated Method for Effective Process Control of Plastic Encapsulated Nichrome PROM's. IEEE International Reliability Physics Symposium, 1977: 143–146.

[42] M. Kitano, et al. Analysis of Package cracking During Solder Reflow. IEEE International Reliability Physics Symposium, 1988, p. 90–95.

[43] I. Fukuzawa, et al. Moisture Resistance Degradations of Plastic LSIs by Reflow Soldering. IEEE International Reliability Physics Symposium, 1985: 192–197.

[44] http://www.jedec.org Oct 10th, 2007.

[45] JEDEC Standard 22 Method A102 Accelerated Moisture Resistance—Unbiased Autoclave.

[46] JEDEC Standard JESD22-A114-B, Electrostatic discharge (ESD) sensitivity testing human body model, JEDEC, 2000.

[47] Ming-Dou Ker, Jeng-Jie Peng, and Hsin-Chin Jiang. ESD Test Methods on Integrated Circuits: An Overview. IEEE International Conference on Electronics, Circuits and Systems (ICECS 2001), 2001: 1011–1014.

[48] Microelectronics Test Method Standard MIL-STD-883D Method 301 5.7, Electrostatic discharge sensitivity classification. US Department of Defense, 1991.

[49] Z.P. Wang. The State-of-the art of Interconnection Technology: Process, Testing, and Design, Keynote paper. In: Proceedings of Third International Conference on Thermal and Mechanical Simulation in (micro)electronics, EuroSIME 2002, Paris, France, 15–17 April 2002, p. 18–26.

[50] Semiconductor Industry Association. The international technology roadmap for semiconductors, http://public.itrs.net, 2005.

[51] G.L. Rosa, S. Rauch, and F. Guarin, New phenomena in device reliability physics of advanced CMOS submicron technologies, IEEE International Reliability Physics Symposium Tutorial, 2001.

[52] Fei Su, Yaofeng Sun, Cheng Yun, Audrey Wong, et al. Development of an Integrated Optical System for Warpage and Hermitage Test of Microdisplay, Optics and Lasers in Engineering Vol. 45 (11), 2007.

[53] Ping An, Yandong He, Yufeng Jin, and Yilong Hao. A Reliability Investigation of MEMS Transducers with Comb Structures. In: Proceedings of the ICEPT 2008, July 28–31, 2008, Shanghai, China.

CHAPTER 10

Prospects for Microsystems Packaging Technology

The microsystems industry, comprised of microelectronics, photoelectronics, MEMS, and some other sectors, is one of the most important industries sustaining rapid growth in the worldwide economy. Design, fabrication, packaging, and testing are the main cornerstones of the microsystems industry. Of them, packaging not only directly impacts the functionality and lifespan of devices and components of a microsystem, it also, to a large extent, dictates the miniaturization, multifunctionality, reliability, and cost of the system. Therefore, it has been drawing increasing attention and is experiencing rapid growth. [1,2]

To meet demand for lighter, smaller, and better-performing system products and the challenges of sustainable development, we should learn from the past and develop new technologies more rapidly. In this chapter, we will discuss packaging materials, packaging processes for microsystems and their applications, as well as impacts on the environment.

10.1   Evolution of Packaging Materials

As the packages of semiconductor devices become increasingly small and thin in size, and also as a result of the increased use of BGA and CSP techniques, the amount of materials used for packaging a device has decreased, posing a dilemma to suppliers of packaging materials. This situation has forced the material suppliers to develop more advanced materials with higher value by using more advanced technologies. Many leading companies of packaging materials have already begun to do so. One of the advances in packaging materials is the improvement of their functionality.

(1) LTCC—The future ceramic packaging material.

Referred to section 3.5 in the book, LTCC is mainly made from glass ceramics and cofired at temperatures of about 900°C. They can be cofired with base metals and have many advantages such as low dielectric constants, low dielectric loss, and passive component integration. In particular, its excellent high-frequency properties make it an ideal material for use in many high-frequency applications. Its application in the military, aerospace, electronics, computer, automotive, and medical industries has been very successful. LTCC is regarded by some experts as the ceramic packaging material of the future.

(2) High thermal conductivity AlN ceramics—The material for packaging power electronics.

Developed in the 1990s, AlN ceramic is one type of high thermal conductivity material used in the production of packaging electronics. Owing to its high thermal conductivity, good insulation performance, CTE matching that of silicon, and low dielectric constant, AlN ceramic is regarded as the ideal material for packaging microelectronics. It has been used in some areas such as microwave power devices, millimeter wave circuits, and high-temperature electronics. As production technology continues to improve and prices continue to drop, AlN ceramics based packaging technology is expected to become the mainstream technique for packaging power electronics.

(3) AlSiC—New type of metal-based composite material.

A1SiC, a metal-based composite material, provides packaging designers with unique material properties suitable for packages demanding superior thermal performance. Compared with traditional packaging materials, A1SiC has the following advantages: first, it can be fabricated in a net shape without complicated postprocessing; second, it features high thermal conductivity, very low density and its CTE matches with Si chip.

As a new metal-based composite material with high performance, A1SiC is expected to replace currently used CuW and CuMo as an important heat-sink material used in electronics packaging in the near future.

10.2   Evolution and Application of Packaging Technologies

The trends for development of microelectronics packaging technology are miniaturization, high density, multifunctionality, and low cost. Generally, it merges into the following: ① Devices are packaged at the chip-scale and are connected with PCB through 2D interconnection; ② The first and second-level packaging processes are integrated into one process and chips are directly installed onto the substrates; ③ Higher density packaging, i.e., packaging more than one chips in one module in either the lateral direction (MCM) or the vertical direction (3D).

The following sections give a brief introduction to the main advanced packaging technologies currently used worldwide.

10.2.1   Ball Grid Array (BGA) Package and Chip-scale Package[3]

Research of BGA started in the 1960s, while its commercial use started only after 1989 when QFP- based SMT was in a bottleneck. From QFP to BGA packaging, line interconnection evolved to 2D planar interconnection, dramatically increasing the I/O pitch of packages, while the sizes and I/O of the packages remaining unchanged. After PBGA was released to the market in the early 1990s, BGA was quickly applied in many fields. There are now a host of BGA packaging and assembly plants in the United States, Europe, Japan, and some other Asian countries. These plants produce BGA packages, mainly used in portable communications products, telecommunications equipment, computer systems, and workstations.

The development of BGA solved the problems that stymied QFP. However, with increasing demand for smaller, more reliable, and more functional microsystems, it could not meet the requirements for higher packaging efficiencies and for approaching the intrinsic transmission rate of chips in developing silicon integration technology. Therefore, in the early 1990s, researchers started developing a superminiature packaging technology called CSP to be used at the chip scale.

CSP is a technology in which the size of package is the same or slightly larger than that of the chip. According to the definition of CSP by the Japan Electronics and Information Technology Industries Association, the ratio of the area of the chip to that of the package should be greater than 80%. The structures of CSP and BGA are the same, and only the diameters of solder balls and the pitch of CSP decrease, making the package smaller and thinner. Thus, more I/Os can be employed in the same package, increasing the density of assembly. CSP can also be thought of as a miniaturized BGA.

CSP provides shorter interconnection and better electrical and thermal performance and is more reliable than QFP and BGA. Because of these advantages, this technique was quickly adopted worldwide. In 1997, it was put into practical use and gradually became the mainstream technique used for packaging integrated circuits with large numbers of I/Os. In Japan, CSP is mainly used in the production of ultra-high-density, miniature consumer electronic products, such as mobile phones, Modems, portable computers, PDAs, camcorders, and digital cameras. In the United States, CSP is mainly used as an alternative to known good die (KDG), which is directly attached to the substrate in a multichip module (MCM) in the production of premium electronic products, or, as a memory device, particularly in a high-performance electronic product with more than 2000 I/O pins.

The typical pitch of BGA is 1.0 mm, 1.27 mm, or 1.5 mm, while that of CSP is usually 0.5 mm, 0.7 mm, or 1.0 mm. Some CSPs with a pitch of 0.3 mm have not yet been used widely. The main reasons include 1 as with QFP at the end of the 1980s, when the pitch of CSP reaches 0.3 mm, the yield of production using SMT assembly is affected; 2 unlike QFP, CSP has its solder joints distributed on a 2D plane and needs to route out the signals of internal solder joints in the PCB. When the pitch is about 0.3 mm, the route-out is almost impossible because the line width/space currently available from proven PCB technology is about 0.1 mm/0.1 mm. Although some companies can now provide line width/space of 75 μm/75 μm, it is still impossible to route out the signal of internal solder joints of CSP with a pitch of 0.3 mm. Therefore, the further decrease of the pitches of BGA/CSP relies on the development of PCB technology.

Before low-cost PCB technologies that could provide line width/space small enough to support CSP with smaller pitches are developed, the development of BGA/CSP will be focused on seeking more cost-effective fabrication methods and better performance of BGA/CSP, including electrical performance, thermal performance, and reliability. This trend is reflected by the appearance of various types of CSPs in recent years. These products include ceramic substrate thin package (CSTP), fine pitch ball-grid array (FPBGA), lead-on-chip CSP, minute mold CSP, land-grid array (LGA), microBGA, bump chip Carrier (BCC), QFN CSP, chip-on-chip CSP, wafer-scale CSP, etc.[5] All or most of the processes of wafer-scale CSPs are completed on wafers on which preceding processes have been completed. Such wafers are eventually diced into separate independent devices. Wafer-scale CSPs have some unique advantages: (1) High-efficiency packaging, process, multiple wafers can be processed simultaneously; (2) Having the same advantages as flip-chip bonding, namely, light, thin, short, and small; (3) Compared with the preceding processes, only the redistribution layer and bumping processes are newly developed in wafer-scale CSPs, and the other processes are all traditional; (4) Eliminating the process of multiple testing in traditional packaging.

For the above reasons, many giant IC packaging houses have invested in R&D and manufacturing of wafer-scale CSPs.

10.2.2   Flip-chip Technology

Among all direct chip attach technologies, flip-chip technology is the most common and typical.

In the flip-chip process, chips can be assembled on high-density FR4 substrates, flexible boards, ceramic printed boards, or glass. No matter what the substrate is, the research has been focused on how to reduce the pitch and enhance the reliability of interconnections.

In the flip-chip technologies using eutectic solder bumps, the minimum pitch in mass production is about 150 μm. In order to further reduce the pitch, Cu bumps were proposed, and some manufacturers are also developing Au/Sn bumps. All these efforts are aimed at controling the flow of bump materials in the reflow process to reduce or prevent bridging of bumps. To enhance reliability, efforts to develop better-quality underfill are being made. Ideal underfill must have good fluidity, have matching thermal expansion coefficients, and cure at low temperatures.

In the flip-chip technologies using bumps made of gold, the minimum pitch could reach about 25 μm. The breakthrough in pitch depends on progress in anisotropic conductive adhesive, nonconductive adhesive, or ultrasonic bonding technologies.

10.2.3   3D Packaging[4,5]

As portable electronic systems become more and more complicated, the need for technologies producing lower power-consumption, lighter and smaller packages for VLSI integrated circuits has been increasingly strong. Likewise, many aviation and military applications are also moving in this direction. To meet these needs, many new 3D packaging technologies have been developed, which stack bare chips or MCMs together vertically. This marks a significant advancement in miniaturization. In addition, shorter interconnection along the Z axis significantly reduces connection resistance, parasitic capacitance, and inductance, leading to a reduction of about 30% in the power capacity of the system.

There are three main methods for 3D packaging. In the first method, components such as R, C, or ICs are imbedded in various types of substrates or multilayer wiring media, and SMC and SMD are then mounted on the surface of the substrate or the wiring media, forming a 3D structure. This method is called the imbedded-type 3D package. Another technique is to configure multilayer interconnections on active substrates in which wafer-scale integration (WSI) has been completed and then to mount SMC and SMD on the top layer to form a 3D structure. This method is called the active substrate 3D package. The third packaging method is based on 2D packaging. It stacks multiple bare chips, packaged chips, multiple-chip modules, and even wafers together to form a 3D structure. This method is called the 3D stacked package. Of these three methods, die-stacking, the so-called "3D packaging technology," was proposed and developed in past several years. The technology integrates the same and/or different functional chips in the vertical direction using 3D interconnection, providing many advantages such as smaller form factor, shorter interconnection, better signal integrity, etc. Therefore, 3D packaging technology has become one of the hottest research topics in the semiconductor industry.

Depending on the types of interconnection technology, there are three typical 3D packages:

(1) Wire Bonding (WB)–Same and/or different functional chips (e.g., flash, DRAM, etc.) are stacked into one package using metal wires (e.g., Au, Al, or Cu) as interconnection medium.

(2) Package-on-Package (PoP)–Same and/or different functional chips (e.g., flash, DRAM, digital, etc.) may be stacked into one package using the mixed interconnection technology, i.e., the top package uses wire bonding technology while the bottom one uses flip-chip technology.

(3) Through-Silicon-Via (TSV)–Same and/or different functional chips (e.g., flash, DRAM, digital, RF, etc.) are stacked into one package using TSV as interconnection medium.

10.2.4   Multichip Module[6]

MCM is a high-density, high-performance, and high-reliability microelectronic product, such as a module, component, subsystem, or system, produced by assembling microcomponents that form electronic circuits on a high-density and multilayer interconnection board using microbonding and packaging technologies. The MCM is a new-generation microelectronics packaging and assembling technique developed, based on PCB and SMT, to meet the needs for, smaller, lighter, thinner, high-speed, high-performance, high-reliability, and low-cost modern electronic systems. It is an important technique for the packaging of integrated circuits.

We usually refer to 2D-MCMs when we talk about MCMs. All components of a 2D-MCM are configured on one plane, although 3D interconnections in the substrate are applied already. As a result of the advancement of microelectronics technologies, the degree of integration of chips has increased significantly. Consequently, the requirements on chip packaging have been increasingly stricter, and the shortcomings of 2D-MCMs have gradually shown up. The current maximum efficiency of 2D-MCM assembly is 85%, which has approached the theoretical upper limit of efficiency of 2D-MCM assembly, becoming a road block to further development of hybrid integrated circuits.

To improve this situation, 3D-MCM was developed. The maximum density of 3D-MCM assembly could reach 200%. In addition to the x-y plane, 3D-MCMs' components are also configured in the vertical direction (the z plane). Compared with 2D-MCMs, 3D-MCMs have the following advantages:

(1) The volume of a 3D-MCM could be reduced to up to 1/10 of that of a 2D-MCM, and the mass of a 3D-MCM could be reduced to 1/6 of that of a 2D-MCM.

(2) The length of interconnection between chips in a 3D-MCM is much shorter than that in a 2D-MCM. This will shorten the delay time of transmitting signals, reduce signal noise and power consumption, and speed up signal transmission/processing.

(3) The 200% assembly efficiency of 3D-MCMs has further increased assembly efficiency and interconnection efficiency, allowing for integration of more functions in a component or system.

(4) Interconnection bandwidth, particularly the bandwidth of storage units, is an important factor affecting the performance of computers and communication systems. An important method for increasing signal width is to reduce delay time and increase data bus width. 3D-MCMs' excellent characteristics meet all these needs.

(5) The number of interconnection I/Os in unit area inside of 3D-MCMs has increased significantly, meaning a higher degree of integration. The increase in the number of internal interconnection I/Os allows for a dramatic decrease in the number of connection points and inserting plates outside of a system, thus improving the reliability of the system.

Despite these advantages, 3D-MCMs also have some disadvantages that need to be overcome. The increase in packaging density will inevitably lead to an increase in the heat produced per unit area of the substrate. Therefore, how to reduce heat is a key issue that must be addressed in the development of 3D-MCMs. The main methods include the use of diamond or chemical vapor deposition diamond films, the water cooling or forced-air cooling process, and the use of thermally conductive adhesives or heat dissipation through-holes. In addition, the 3D-MCM, as a new technique, needs to be improved gradually; equipment for fabricating 3D-MCMs needs to be upgraded, and new fabrication software needs to be developed.

With its unique advantages in assembly density, packaging efficiency, signal transmission speed, electrical properties, and reliability, MCMs are currently the most effective way of maximizing integration degree of systems, establishing high-speed electronic systems, and achieving the miniaturization of systems with multiple functions, high reliability, and high performance. MCMs started their applications back in the early 1980s. However, the applications were limited to the military, aerospace, and mainframe computer sectors because of their high costs. Along with the advancement of technologies and the reduction in costs, MCMs have found increasingly wide applications in electronic products such as computers, communication equipment, radars, data processing equipment, automobiles, industrial equipment, instruments, and medical devices. It is the most promising advanced microsystems assembly technology.

10.2.5   System in Package[5,7]

There are two ways of achieving the highly integrated functions of an electronic system. One is system on chip (SOC), namely, building an electronic system on a single chip, and another way is System in package (SIP) or system on package (SOP), namely, achieving the functions of a system by means of packaging. Academically, they are two technical routes, and just like the relationship between monolithic integrated circuits and hybrid integrated circuits, they each have their own advantages and application markets. Their relationship is complementary both technically and in application. SOC is mainly used in high-performance products with a long lifespan, while SIP is mainly used in consumer products with a short lifespan.

SIP is a process in which various types of circuits such as CMOS, GaAs, and SiGe, or optical electronic devices, MEMS devices, or various types of passive components such as capacitors or inductors are integrated in one package to establish a system using proven assembly and interconnection technologies. In the past several years, SIP solutions have been widely adopted by the electronic packaging industry for different applications such as cellular phones, base stations, PDAs, MP3 players, camcorders, digital video recorders, digital cameras, laptop computers, PCs, internet routers/switches, servers/workstations, etc.

The main benefits of SIP include (1) low manufacturing cost because of the use of currently available commercial components; (2) short time period required for bringing products into the market; (3) great design and technological flexibility; (4) relative ease of integrating different types of circuits and components. The single integrated module (SLIM) developed by the Georgia Institute of Technology is a typical SIP. This technology increased the efficiency, performance, and reliability of packaging by 10 times, while the size and cost of packaging dropped significantly. By 2010, the wiring density of the technology is expected to reach 6000 cm/cm^2, the heat density is expected to reach 100 W/cm^2, the component density is expected to reach 5000/cm^2, and the I/O density to reach 3000/cm^2.

10.2.6   New Areas for Packaging Technology Development

1. MEMS Packaging

MEMS processing technologies have improved considerably in terms of speed, range of application, and multifunctionality. Broadly speaking, a microsystem consists of MEMS, signal conversion, processing units, and electrical and mechanical packaging. After years of rapid growth, MEMS technologies have been widely used in the automobile, medical, and communications industries, as well as some consumer electronic sectors. It is anticipated that the MEMS market will grow at an even faster pace.

However, packaging technologies have posed considerable restraints on the development of MEMS products. Like any other semiconductor device, MEMS devices also need specialized packaging to protect them from harsh environments, to provide electrical signal connections and mechanical support, and to dissipate heat generated.

In MEMS packaging, delicate chips or actuators must directly contact with the working media, which are often very harmful to chips. In many cases, inert gases or vacuums are required inside the MEMS packaging. In addition, almost all microsystem packaging contains small but complicated 3D structures.

Many single-chip ceramics, molding, chip-scale packages, and wafer-scale packages have been successfully used in MEMS. MEMS multichip and 3D packaging technologies are currently still under development.

2. Optoelectronics Packaging

When many aspects of the electronics industry started declining, the optoelectronics industry began to emerge as a new bright spot. The growth of the optical communications market has provided both new opportunities and challenges for EMS suppliers. Optoelectronic devices are the combination of optical components and electronic circuits, consisting of active components, passive components, and the interconnections forming optical channels. Optoelectronics packaging is used to integrate these optical components and electronic packages to form a new module, which can be regarded as a special multichip module with fewer I/Os and smaller size of chips.

The main problems faced by optoelectronics packaging are the mismatch between the high digital signal transmission speed and the low optical signal conversion rate, as well as the integration of optical functional components. Since a complicated substrate may be used in packaging, it is very important to understand the properties of the material. Optoelectronics packaging requires complicated designs. It must contain optical, electrical, thermal, and mechanical designs required by a system. Thermal design is particularly necessary because optoelectronic devices may be sensitive to working parameters such as wavelength. Temperature often needs to be adjusted in order to enable optical components to work under a specific wavelength. To adjust temperature, a cooler with appropriate heat reduction functions is needed in the packages. In the future, the packaging density of optoelectronic components will be higher and these components' working speed will be faster. This will make the heat dissipation of optoelectronics packages more prominent. Optoelectronics packaging plays an important role in the optoelectronics industry.

3. Wide-bandgap Semiconductor High-temperature Electronics Packaging

In recent years, research and development into wide-bandgap semiconductor devices, made from high-temperature semiconductors such as SiC, GaN, AlN, and semiconductor diamond, have attracted much attention. The hightemperature semiconductors used in the production of these devices feature wide forbidden band, high breakdown voltage, large heat conductivity, high carrier mobility, low dielectric constant, and high radiation resistance and are regarded as the third generation semiconductors after Si, GaAs, and InP. These materials have bright prospects for use in high-temperature, high-power, and high-frequency electronics and short-wavelength optoelectronics. Wide-bandgap semiconductor devices are usually used in challenging environments, for example, under high temperatures, and therefore they need special packaging. In the past, little attention has been paid to the packaging of devices used under high temperatures. With the evolution of high-temperature–resistant electronics, the importance of packaging has been increasingly recognized. GaN devices, for example, can work under temperatures above 600°C, an environment where conventional electronics packaging materials such as fiberglass epoxy circuit boards, copper-coated wires, and tin-lead solder are no longer applicable. Even standard alumina ceramics packaging cannot be used in temperatures above 300°C, and therefore there is an urgent need for new packaging materials and technologies.

The key to high-temperature–resistant electronics packaging is not to find materials that can withstand, high temperatures, but to find materials that are compatible with assembly techniques. Thermal conductivity, thermal expansion coefficient, oxidizability, diffusion, and some other factors, for example, have become the key to high-temperature–resistant electronics and are key considerations in selecting materials. The ideal material currently available for high temperature packaging is aluminum nitride. However, some problems such as metallization and airtight sealing under high temperatures must be solved before actual implementation.

4. RF Packaging[8]

In recent years, the wireless communications market has witnessed explosive growth, leading to a surge in millimeter wave applications, including Local Multi-point Distribution System (28 GHz), Wireless Local Area Network (60 GHz) and automotive Millimeter wave anticollision radars (77 GHz). Developing these technologies requires low-cost, miniaturized, and large-capacity millimeter wave packages urgently. The current constraints on the working frequencies of the above products are the parasitic parameters of their packages rather than the chips of integrated circuits. These parasitic parameters–physical, distributed, and electromagnetic field parameters–greatly impair the frequency response of devices as well as the integrity of signal. Packaging is the real root cause for limited transmission speed. The higher the working frequency, the greater the impact of packaging.

5. Micro-opto-electro-mechanical System (MOEMS) Packaging

MOEMS is a new type of technology comprised of micromechanical optical modulators, micromechanical optical switches, IC, and other components. It is a new application of MEMS technology in optoelectronics. It realizes the seamless integration of optical devices and electrical devices by taking full advantage of the benefits of MEMS, such as small sizes, diversity, and microelectronics properties. In recent years, the application of optical MEMS devices in the communications industry has drawn much attention (particularly in the optical network and optical switch areas). Since optical MEMS devices heavily rely on high-precision optical, electrical, and mechanical designs, they have some special requirements for packaging. In addition to complete circuits for light and electricity, optical MEMS packaging must provide some other properties such as air tightness, mechanical strength, dimensional stability, and long-term reliability. In addition, the increase in I/O pin counts requires higher density of substrates and packages and different applications require different packaging. Packaging is a key process in the manufacturing of MOEMS, accounting for 75%−95% of the total cost.

10.3   Evolution of Packaging Technologies and Environmental Protection

Many people believe that the high-tech industry is environmentally friendly; but in fact, it is not. Statistics show that only 7% of raw materials used in the processing of high-tech products are used in end products, 80% of products are scrapped after being used once, and 99% of raw materials are junked within six weeks after their first use. In contrast, 80% of the production costs are determined by designers. Therefore, it is meaningless to focus all attention on technical research when developing microsystems and microsystems packaging technologies since we still have to pay attention to the impact of these technologies on the environment and resources. Not all technologies that bring temporary convenience to people are necessary, and maybe it is correct to say that only sustainable technologies and industries can truly benefit people. While bringing convenience to people, high technologies have also created a large amount of modern garbage–electronic wastes. Statistics show that the volume electronic wastes is growing three times faster than that of household garbage.

Electronic wastes are also a serious social problem that needs to be confronted.

10.3.1   Electronic Waste Pollution

Electronic wastes are currently the world's fastest growing type of garbage and have become a new public hazard. Electronic wastes refer to all waste electronic products including electrical household appliances, IT equipment, communications equipment, TV sets, audio devices, lighting equipment, monitoring equipment, electronic toys, and motor driven tools. Of all the electronic wastes worldwide, 80% are exported to Asia and 90% of Asia's imports eventually find their way to China.

Therefore, China takes 70% of the world's all electronic wastes annually. China itself also produces hundreds of millions of waste electrical appliances each year. For China, effectively disposing of electronic wastes to protect the environment is a very urgent task.

The main hazardous substances in electronic wastes include lead, mercury, cadmium, hexavalent chrome, polybrominated diphenyl, and polybrominated diphenyl ether. Waste electronic products are junked carelessly or recycled incorrectly. Their harm to the environment is no less than that by conventional garbage. If improperly treated, electronic wastes will create serious pollution. Untreated landfills, burning, or throwing away will all produce hazardous substances that can heavily pollute soil, air, or water sources.

As the information technology industry grows faster and faster, products are also being phased out at a faster speed. Those products that were updated every three to five years in the past are now being updated about every 18 months. The latest statistics show that the replacement rate of high-end mobile phones in China is about every six months.

10.3.2 Control at Workshops

In the face of various hazards caused by electronic wastes, environmentalists have called on manufacturers to take account of how to recycle their products in an environmentally responsible way when they design the products. The European Union has stipulated that starting on July 1, 2006, no electronic products containing hazardous metals such as lead are allowed to be sold in the EU. Therefore, sustainable development issues must be taken into account when selecting materials.

1. Lead-free Solders

Sn-Pb solder has been used to connect metals for a long time. Its first use can be dated back to 2000 years ago. Sn-Pb solder and its alloys have many advantages, such as easy operation, low melting point, excellent operability and plasticity, and good wettability to Cu and its alloys. Today, soldering has become an irreplaceable technique used in the interconnection and packaging of electronic devices and circuits.

Lead-based solders, particularly eutectic and hypoeutectic Sn-Pb solders, are widely used in modern electronics and circuit packages. However, environmental and human health concerns arising from the toxicity of lead and legislation limiting the use of lead-based solders have prompted manufacturers to quicken the pace of finding lead-free substitutes for lead-based solders in the production of electronic products.

The industrial sector has developed several types of lead-free solders with performance similar to that of eutectic Sn-Pb solder. Among these new products, a tin-silver-copper solder is the most widely accepted and has been adopted by a host of companies.

2. New Materials for Printed Wiring Board Packaging

In recent years, in order to meet the requirements of environmental protection, people have stepped up their efforts to develop and popularize lead-free solders and halogen-free and antimony-free materials. To reduce the burden on the environment, future materials will be required not to produce endocrine disrupting chemicals, carcinogenic substances, dioxin, or furan when burnt.

Halogen-free materials currently being sold on the market are mainly used in the production of general-purpose printed wiring boards. In producing halogen-free flame-retardant printed wiring boards, using reactive organophosphorus, metal hydrooxide flame-retardants, or underfill composed of different oxides can produce reliable electrical properties and meet the requirements of the UL 94-V0 flame retardation specifications.

However, it is very difficult to achieve excellent flame retarding properties and physical properties of resin simultaneously. We urgently need to develop flame retardants that have less impact on the environment as the next generation product. Since the demand for epoxy resin is very large, people are accelerating the development of technologies for recycling printed wiring boards, and the technologies for recycling such materials including fiberglass are close to practical use.

3. Conductive Wires Used in Electronics Packaging

Wire bonding material used in electronics packaging is one of four basic structural materials used in the production of electronic products. Along with the boom of the electronics industry, wire bonding material is set to see major development and will in turn give a great boost to the electronics industry. In Ultra-Large-Scale Integration Circuit wire bonding, gold wire is the most common conductive wire used. In order to reduce the cost of packaging, manufacturers have always been trying to find a less expensive material as a substitute for expensive gold wire. Using copper wire could not only reduce the cost of producing devices but could also enhance the strength of connection compared with gold wire. The use of copper wire has provided an impetus for the development of low-cost, small-pitch, and high-I/O pin count device packaging.

To reduce the diameter of gold wire used in bonding, improve connectivity, and cut costs, Japan's Tanaka Electronics and Sumitomo Metal Mining and Germany's Heraeus Group have developed a series of new types of high-performance gold alloy wires, including Au-Ag, Au, Ni, and Au-Cu and have now begun to put them into practical use.

4. Environmentally Responsible Photoelectronics Assembling and Bonding Methods

Photoelectronics packaging performs three functions–providing mechanical holding, enhancing environmental tolerance, and providing electrical connections with other components. One of the environmentally responsible photoelectronics assembling and bonding methods is as follows: (1) pure nitrogen/vacuum/nitrogen; (2) vacuum baking/vacuum heating, 250°C, 2 hours; (3) aspirator activated, 400°C, 15 minutes; (4) lifted; (5) seal bonding, 170°C; (6) Maintain the bonding temperature for 30−40 seconds; (7) cooling under nitrogen.

This method needs no soldering flux and is therefore environmentally friendly.

5. New Types of Passivation Materials

SiC is a novel passivation material used in microsystems. The Plasma-enhanced chemical vapor deposition process is used to produce noncrystalline SiC film. In the past, silicane and methane were used in the production of carborundum; however, these gases are harmful to human health. Later, a German solid state technology research institute began to use hexamethyl disilazane (HMDS) as the liquid particle in the process. HMDS is nontoxic and harmless, and its deposition speed is 10 times that of silicane. In addition, the deposition temperature of HMDS is relatively low, only 200−300°C, making it possible for HMDS to deposit on many types of substrates and pyrex glass. SiC formed on the substrate can resist the corrosion by KOH or HF. Therefore, HMDS is a relatively ideal passivation material.

6. Cleaning Technology

To eliminate the pollution to the environment occurring in the cleaning process, a design-for-environment (DFE) electronic circuit board cleaning agent has been designed. This design is for batch water rinsing or online spray rinsing. The agent can remove surface-mounted modules, rosin-based soldering flux on advanced packages, residual soldering flux left after the use of no-clean techniques, water-soluble soldering flux, unsolidified soldering paste, or residual products of modules.

DFE SMT water-rinsing agents meet the following requirements for performance: being free of Ozone Depleting Substances; having a volatile organic compond (VOC) content of lower than 25 g/L; being free of hazardous substance; being compatible with electronic circuit components; being able to remove ion pollutants; and meeting consumers' expectations on costs.

10.3.3   Electronic Waste Treatment

A worldwide well-known method for treating electronic wastes is the Finland model. In Finland, people can take their waste electrical household appliances to either stores or appliance recycling centers for recycling when they decide to buy new appliances and electronic products.

In Finland, local environmental protection and garbage treatment departments send specialized vehicles to recycle electronic wastes in residential areas regularly. In Helsinki, there are currently three garbage stations and one garbage recycling center responsible for recycling electronic wastes from households and enterprises. Garbage treatment departments in Helsinki also send vehicles twice each year to the downtown area to recycle electronic wastes. People need to pay a treatment fee for the recycling of their electronic wastes. The cost of treating one ton of electronic wastes currently in Finland is around $700-1300$ euros. Metals, plastics, and glass recycled from electronic wastes are mainly reused within Finland. Aluminum, iron, copper, and zinc are the raw materials used most efficiently among all recycled materials. By 2006, the amount of electronic wastes recycled across Finland was expected to reach 80,000, tons and by 2010 the recycling rate of electronic wastes in the country was expected to reach nearly 100%.

The U.S. Environmental Protection Agency has suggested some ideas and approaches for recycling electronic wastes. For example, it proposes waste electronic products could be donated to schools or other nonprofit organizations; enterprises or individuals could choose to buy recycled products when purchasing electronic products; electronic product owners could sign contracts with their suppliers to lease their waste electronic products; and manufacturers could consider designing electronic products that are easier to upgrade or reassemble.

Starting with the formulation of new regulations, Germans have taken a wide range of measures to recycle as many waste electrical household appliances as possible and then treat them after classifying them. Doing so could not only greatly reduce the pollution to the environment by waste electrical appliances but could also enable people to make full use of the useful components of these appliances, contributing to the conservation of energy and resources.

Some electrical appliance suppliers are trying a new selling mode, in which customers need to make an additional payment for the future recycling of the products they are going to purchase.

10.4   Concluding Remarks

Electronics packaging experienced unprecedented changes in the 1990s. From the through hole method of the 1970s to today's 3D packaging, electronics packaging technologies have been improving continuously. The evolution of electronics packaging technologies has had a far-reaching influence on the development of advanced electronic equipment and even on peoples' lives as a whole. Today, the packaging industry has ridden on a fourth wave of development–the system-level packaging. Packaging technologies are facing not only challenges but also great opportunities for further development.

Achieving sustainable development is a very urgent task for China, one of the world's major manufacturers. To an economist, stimulating the economy advances human civilization, and therefore economic development is a top priority. However, humans, the most intelligent creatures on Earth, must be aware that the economy is of little importance compared to the ecological system. Should technologists take time to consider this fundamental question of the health of the ecology? Looking at this question from someone else's perspective may lead to the realization that today's rapid economic and technological growth has been achieved at the cost of development opportunities for future generations. Maybe we should shift our focus from stimulating economic growth with technology to those issues instrumental to sustainable growth and try to find the golden ratio, a balance between technological development and development that is sustainable to human beings.

Questions

(1) With the increasingly serious environmental problems, what should an engineer consider in microsystems design?

(2) In the next few decades, in what areas is microsystems packaging expected to see major development?

References

[1] Shangtong Gao and Kewu Yang. The Advanced Microelectronics Packaging Technology. Electronics and Packaging, 1 (2004): 10–15.

[2] Shuping Zhang and Hongyu Zheng. New Progress of Electronic Packaging Technology. Electronics and Packaging, 1 (2004): 3–9.

[3] Fei Xian. The Advanced Chip Package Technology. Electronics and Packaging, 4 (2004): 13–16.

[4] Jinqi He. Three Dimensional (3-D) Packaging Technology. Microelectronic Technology, 4 (2001): 1–10.

[5] Yole Développement. 3D IC and TSV report, Yole Development, Nov., 2007.

[6] Yun Zeng, Min Yan, and Xiaoyun Wei. Multichip module technology. Semiconductor Technology, 6 (2004): 2–4.

[7] Tummala, R.R. Packaging: Past, Present and Future. 6th International Conference on Electronic Packaging Technology, (2005): 3–7.

[8] Katsuaki Suganuma. The Actuality of Microelectronics Packaging for Wireless Conmmunication. Equipment for Electronic Products Manufacturing, 12 (2004): 3–4.

Index

Printed and bound by CPI Group (UK) Ltd, Croydon, CR0 4YY

21/10/2024

01777046-0004